CONTRACTING TO BUILD
YOUR HOME

CONTRACTIN

*How To Avoid Turning
The American Dream
Into A Nightmare*

G TO BUILD
YOUR
HOME

HERSCHEL G. NANCE

Sunstone Press
Santa Fe, New Mexico

First Edition
Printed in the United States of America

Library of Congress Cataloging in Publication Data:

Nance, Herschel G., 1922-
   Contracting to build your home : how to avoid turning the American
dream into a nightmare / Herschel G. Nance. -- 1st ed.
    p. cm.
   Includes index.
   ISBN 0-86534-160-5 : $12.95
   1. House construction. 2. Contractors. 3. House buying.
 I. Title.
 TH4811.N249
 690' .837--dc20   91-32694
                                  CIP

Published in 1992 by SUNSTONE PRESS
     Post Office Box 2321
    Santa Fe, NM 87504-2321 / USA

CONTENTS

PREFACE

Purchasing a home or contracting for the building of one is one of the most important financial transactions that an individual will make during his lifetime. Therefore it is important that he approach the transaction with the greatest caution. Many people buy or build homes without proper investigation into the matter and as a result, either lose substantial sums of money or are forced to accept homes that are not even remotely similar to structures they had in mind when they first started considering a new home. The home building profession abounds with general contractors who are only marginal in their skills or are intent upon victimizing the client. Courts have traditionally decreed that contractor misdeeds are civil in nature and not criminal. Hence,when one employs the wrong builder, there is little that can be done to alter the course of events which have been set in motion. Often, the client is doomed to accept an inferior product. A person should do everything possible to assure that he has acquired a builder who is of the highest possible integrity and with the necessary talent to construct the home that the client has commissioned him to build. If one selects the right builder and follows the directions outlined in this book, the buying or building of a home can be an enjoyable experience, not a nightmare.

ACKNOWLEDGMENTS

The author wishes to express his thanks to the numerous people with whom he served for twenty-two years in the US Army Corps of Engineers, and for the training and opportunities which the corps provided for the author to inspect construction and to become adept at the skill. The experiences gained enabled the author to identify areas of concern in his own house which he contracted to build and to set down the fundamentals that everyone should follow when they intend to buy or build. The author is especially indebted to his wife Reba, his daughters Nancy and Sandra and son, John, all of whom urged him to write this book and to provide the benefit of his many years of training which had thus far never been made available to the general public.

WARNING / DISCLAIMER

This book is designed to provide guidelines in regard to the subject matter covered. It is sold with the understanding that the publisher and author are not engaged in legal, accounting, or other professional services. If legal or other professional services are required, a qualified professional should be engaged. It is not the purpose of this book to provide answers to all questions that may arise when one contracts for the building of a home, and in no sense, is it to take the place of, or preclude the necessity for competent legal advice. This book is written only with the hope of the author that your reading and understanding the contents will alert you to questions that you should ask and pitfalls which you should attempt to avoid when contracting to build your home. Most of all, it is hoped that the contents of the book will help you to recognize the many subjects for you to address with your attorney. There are many books in print which treat many of the individual subjects in this book in greater depth. Depending upon your needs, you are urged to read more of the available literature which may go into greater depth on the subject which you choose to address, and to tailor the information to suit your needs.

Maximum effort has been made to make this book as accurate and as informative as possible. However, there may be mistakes both typographical and in content. Therefore, you should use this book only as a general guide and not as the final source of information that you should have when contracting to build a home.

The purpose of this book is to provide elementary information and to entertain the reader. The author and the publisher shall have neither liability nor responsibility to any person or entity with respect to any loss or damage caused, or alleged to be caused, directly or indirectly by the information contained in this book.

*Contracting To Build
Your Home*

ONE: BEFORE
YOU START

This book provides guidelines for those who are buying or contracting for the building of a new home. Adhering to the principles contained herein will not only insure a quality home; it can save you thousands of dollars. These savings and increased quality are not limited to the prices paid for the house or to the costs of incorporated labor and materials. The book identifies a number of ways that the owner can participate directly in the construction process, either by providing much of his own labor, doing his own subcontracting, or sharing designated responsibilities with the general contractor.

Psychologists say that all of us have four basic psychological needs: status, affection, independence, and security. They claim that these needs working together or in isolation dictate many of the ways we adjust to our environment. In our desire to own the house in which we live, most of us are responding to at least one and sometimes as many as three of those basic needs. In acquiring our homes, we Americans are responding in some degree to our needs for status, independence, and security, and the degree to which we respond to each need defines the varying tastes that we have for the homes that we buy or build. In today's mobile society, it is not uncommon for one individual to own as many as ten or twelve different homes in succession as his work place changes from state to state. During the early years of married life, it is not always possible for a young couple to attain the home of their dreams, but almost invariably, that couple will gradually develop their own concept of a particular type home that they plan some day to buy or build.

When the day arrives for a couple to buy or build their dream home, numerous questions arise as to the procedures they should follow in quest of their goal. These questions are some of the most important issues that a couple will face during their lifetime. Therefore, it is prudent that the many and varied tasks associated with attainment of the goal be undertaken only after the parties have become familiar with potential problems and the safest and most economical approach for completing each task. This book provides answers to questions of both procedure and economics associated with the process and, if followed, has the potential to save the client enormous sums of money, while simultaneously insuring that his new home will be of the highest possible quality.

The primary objective of this book is to define the pitfalls associated with contracting for the building of a home, to outline the fundamentals that one must know before contracting to buy or build a home, and to provide specific step-by-step guidance for the individual once he has made the decision. The objective is attained by dividing the principal fundamentals into fourteen chapters, which address specific aspects of the decision to buy or build and/or pitfalls to be avoided while the process transitions from concept to completed home. Outlined below is a summary of each chapter with a brief description of the importance of the subject which the chapter addresses.

Chapter 1 introduces the rest of the book, and tells why it is essential for one contracting for the building of a home to follow the fundamental guidance outlined in succeeding chapters.

Chapter 2, Selecting an Architect, a Lot, and a Builder, provides detailed information to assist in selecting an architect , the proper lot and a reputable builder, along with a number of admonitions against factors not commonly known by people outside the building profession.

Chapter 3, Fixed Price or Cost-Plus-Fixed-Fee, identifies the two main types of contracts that are regularly employed when one engages a professional builder. It identifies the Cost-Plus-Fixed-Fee contract as an instrument containing greater risk than the fixed price contract, and identifies the risks involved. It points to the fact that a fixed price contract is the route taken by a majority of clients, but identifies a number of specific actions that the client will need to take prior to negotiating the contract.

Chapter 4, Purchasing a New House, identifies the danger areas for which one should look if he decides to purchase a new ready-built house, and compares the costs and quality of new ready-built houses with those built under contract with a builder.

Chapter 5, Basic House Construction Terms, provides a list of the basic terms used in the construction process and discusses ways to assure that a general contractor is performing the work properly. Familiarity with basic construction terms is a necessity for one buying a home, subcontracting his own home, or employing a general contractor.

Chapter 6, Subcontracting Your Own Home, furnishes detailed instructions for one who is willing to manage his own subcontracting, pointing out that the process is not nearly as difficult as imagined. This chapter emphasizes that, whether one employs a builder, or does his own subcontracting, he will need to be familiar with the subcontracting procedures and how they may affect him as the client.

Chapter 7, The Beginning of the Nightmare, is the first of two chapters designed to enhance the other twelve chapters by narrating step by step the unfortunate experiences of the author when he elected to contract for the building of a home without first acquainting himself with potential pitfalls in the endeavor and becoming familiar with the sequence of events involved in contracting for the building of a home, as well as the relationship one will have with contractors, subcontractors, suppliers, lending institutions, and attorneys.

Chapter 8, Obtaining Mortgage Money, instructs the reader on the various types of mortgage loans that he may obtain, both for construction loans and permanent loans, and provides guidance on how to find the best rates and terms.

Chapter 9, Doing Much of Your Own Construction, points to many ways that one can realize considerable savings by doing much of his own construction work, even though he may not have substantive experience in the areas concerned. The chapter provides guidelines for obtaining a general contractor who will agree to self help by the client, and high lights the method and advantages of inserting self help details into the contract.

Chapter 10, The Building Inspector, describes the role that the county or city Building Inspector performs in the building of a house, and elaborates on the ways that the inspector can assist one doing his own subcontracting, as well as the service the inspector provides in assuring that a general contractor performs quality work.

Chapter 11, Twenty Six Tips for Saving Money, offers twenty-six options that can save a client a considerable amount of money. Some of the options apply to substitutions in material for the home; others apply to work that the client may do for himself.

Chapter 12, Awaking from the Nightmare, is the second of two chapters which narrate the personal experiences of the author and his wife when they entered the house contracting arena without having prepared themselves to cope with the numerous adversities which they were almost bound to encounter. This chapter tells how they eventually extracted themselves from the nightmare in which they had been mired for two full years.

Chapter 13, Preparing the Contract, advises that the contract should be prepared by an attorney, but not before the client has provided the attorney with a list of the many things that he wants incorporated into the contract. Formulating the contract calls for specificity in every detail, including statements of what,when, and where things are to be done, along with precise statements of penalties for failure to meet milestones. The contract must also state how changes are to be negotiated, and it should define the conditions under which either party can terminate the contract. The chapter provides detailed guidelines for the client to follow in formulating the list of items for his attorney to put into the contract.

Chapter 14, Avoiding the Pitfalls provides a summary of the most important elements in each chapter and describes the selection of a good builder as the singularly most important step in contracting for the building of a home. In addition, it describes the general characteristics of a poorly qualified contract seeker and advises the reader to be especially alert for such a candidate so as to eliminate him from consideration in the earliest possible stages of consideration of possible candidates for award of the building contract.

The lessons contained in the above identified chapters acquaints those thinking of contracting for buying or building a home with numerous options to save money; tells them how to play a major role in the construction process, and identifies numerous ways for them to enhance the quality of the finished product, and avoid the pitfalls associated with some contracts.

There are many reputable builders who are professional in every sense of the word and execute their contracts with a view toward providing their clients with the maximum amount of quality workmanship for their money. However, like all professions,the building profession has its share of unethical contractors. For anybody about to contract for building a home, an important first step is to acquire the necessary knowledge to discriminate between unethical builders and those who are trustworthy. The definitions,instructive matter,and procedural outlines contained in this book provide the necessary information to assure that the clients select competent builders. Following these guidelines throughout the construction process will insure that the completed home is exactly the product which the client initially desired.

Seventy-five percent of this book is dedicated to the technical fundamentals of things that one should do or should not do when contracting for the building of a home. The message would not be complete, however, without examples of what can happen to one who fails to heed the lessons contained in this book. A lesson is always learned easier and · retained longer if it can be applied to real-life situations. An entire set of real life situations reflecting the experiences of the author is related as the book progresses from one set of instructions to the next. It is not only interesting from a human interest point of view; it vividly portrays the things that can happen to you and every other individual if you fail to exercise the necessary precautions as you enter the arena of contracting for the building of a home.

Each day the news media reports flagrant cases of con artists operating throughout the country. We have all read accounts of their activities and we have observed that, in general, they tend to have certain common characteristics, foremost of which is the con artists developing trust in their unwary victims whom they betray to achieve their objectives. There have been a wide variety of these cases reported, but few reports have dealt with the theft and greed that is paramount in the building profession.

A random sampling of only two developments in one county in one of the most populous southern states reveals that out of sixteen cases with sixteen different builders, six of sixteen clients experienced extraordinary financial loss as a result of misdeeds or outright fraud at the hands of builders. In many cases, the sums of money lost by unwary clients were enormous. All this occurred when the only error of the clients was their having been trusting enough to award house building contracts to the wrong builders.

The legal systems of most states tend to encourage the practices of the infamous builders by decreeing that the money that they steal or their cheating on materials or workmanship is a civil, not a criminal offense. This aspect of the legal system encourages builders who are so inclined to exercise every trick of the trade to attain unearned gains at the client's expense. In many cases these gains are just as nefarious as those attained by the criminal who holds a pistol in the ribs of a victim while committing the crime of robbery, rape, or murder.

The building profession is so loosely policed by our legal system that it offers an alluring goal for skilled and unskilled laborers in the profession, the most clever of whom are quick to recognize the opportunities for personal aggrandizement. In most states, a builder doesn't even have to be licensed to enter into a contract where sums of less than $40,000 are involved. If a would-be builder is looking for higher game, he is required only to pass a simple test containing questions from subject matter in a small code book, generally not more than seventy-five pages or so in length and capable of being memorized overnight by an astute individual.

Once an individual obtains his builder's license, a hundred different avenues become available for him legally to cheat the uninformed public. Among them are the purchase and incorporation of sub-standard materials. Under the cost plus fixed fee contract, avenues are use of the client's money for purposes other than building the client's house; failing to pay vendors and subcontractors and thereby exposing the completed house to the possible imposition of liens by unpaid creditors, and covering defects with wallboard and paint. It is not uncommon for cost plus fixed fee contractors to purchase materials for one house, but have them delivered elsewhere, or to permit subcontractors to haul away material purchased for the client's building. There are numerous other methods and techniques employed by builders to achieve their ends, depending upon whether the builder has a fixed price contract or a cost plus fixed fee contract. Unethical practices sometimes used by contractors to exploit their clients will be discussed within the content of this book.

It has been my own experience to have been victimized by the type of builder just described. The experiences that I encountered during the two year ordeal almost defy belief. In this book, I have related the events as they occurred from the day that my wife and I decided to contract for the building of our home, until it was finished some fourteen months later. This encompassed a period wherein we were victimized not only by the builder whom we hired to build our home, but also by two attorneys whom we hired to set things straight. It is a sordid story, one that has to be lived if it is to be believed, for much of it tends to refute the principles that we learned in early civics classes. It certainly makes one think twice before endorsing the view that in our country's legal system, justice will prevail.

Succeeding chapters of this book perform a two fold objective: to set forth the guidelines for contracting for the buying or building of a home, and to provide for your learning reinforcement by narrating a chronological sequence of events that occurred when two people set out on the road to contract for building a home without first identifying the pitfalls and heeding the fundamental guidelines so vital to such an effort.

TWO: SELECTING AN ARCHITECT, A LOT, AND A BUILDER

THE ARCHITECT: The selection of an architect to design your home is equally as important as the selection of a lot or a builder. Many who have travelled the road of having a home built argue that the architect is more important than any other professional involved in the process. The architect is not only the designer, but an advisor who can serve you during all phases of construction. Most architects can be engaged to perform as much or as little of a list of basic services as you feel is necessary. The basic services include design, preparing all construction documents, managing the bidding process, and administering the contract. It is possible that you may employ him for only the first service, or for each additional service in the order listed. Architects not only offer a list of basic services, many offer a number of options from which you can choose, depending upon how definite you are in regard to various aspects of the home you want constructed. Today, there are architects in each city who have a library of various plans for you to review to identify the plan which most closely suits your taste. Once you have identified the home with the style and configuration of your choice, he agrees to modify the plans as necessary to bring them into complete conformity with your desires. He earns his profit from the fees that he charges to modify the plans. If you don't need much modification, your architectural fee will be low. If you require numerous modifications, your fee will be higher. In other cases, people have their own idea as to what they want in a house and they are not willing to accept any existing plan, modified or otherwise. In such cases, it is necessary for the client to describe his visualization in thorough and descriptive terms, so the architect can design the home from scratch. A design of this nature is necessarily more expensive than any other, because it calls upon the architect to make accurate perceptions of what is desired and for him to stay at the drawing board changing his design as many times as necessary to transform your ideas into completed plans and specifications. If you can afford the cost, it is wise to employ the architect for all four of the basic services, thereby relieving you of much of the hassle of conveying your thoughts and obtaining agreement with the general contractor and subcontractors.

In many localities, the building codes require that residences be designed by architects. In others, there is no such stipulation. In the latter case, there are sometimes general draftsmen who have become experienced in designing homes and are available for design work ranging from small renovations and additions to full scale homes of various descriptions. These draftsmen can generally be retained for smaller fees than qualified architects, but you are urged to refrain from engaging them unless you know one who is thoroughly experienced and has a long record of successful designs and pleased clients. When you start your search for an architect, there are a number of factors that you should consider.

1. Does he have an architectural degree and has he designed a number of houses in your area?
2. Does he quote a fee that is within your budget?
3. Is he recommended by previous clients?
4. Can he recommend reliable contractors and/or subcontractors?
5. Does he show enthusiasm for your project?
6. Does he have a number of clients, yet not too many to keep him from devoting the proper time to your project?
7. How well did his estimate on some of his previous jobs coincide with actual costs when the jobs were finished?
8. Can he show a record of having successfully designed buildings of the same general architectural style that you are seeking?
9. If he recommended a contractor from one of his former projects, was the house built with quality workmanship?
10. Is he willing to enter into a contract with you whereby you have the right to terminate his services if for any reason you become dissatisfied with his work.
11. Does he work at an hourly rate or for a flat fee? You should attempt to obtain him at a flat fee if possible; otherwise, it may be difficult for you to budget for the work.

Keep in mind that your architect is probably a busy man. If he isn't, you probably wouldn't be interested in retaining his services. Because he is busy, his time is limited. If you haven't been able to make up your mind, or you are constantly requesting changes, the architect may lose interest in your project, and his creative effort may be affected accordingly.

THE LOT: A frequently stated adage in the real estate profession says that the most important three factors in selecting property are location,location,and location. This adage is certainly valid when it comes to choosing a lot on which to build a home, for location is the greatest determinant of resale value of the property as well as the marketability. Outlined below are the pluses and minuses one should consider when searching for a lot.

Generally, it is desirable to locate a house within a development or an area where other homes in the same general price range have been or will be built. There is a tendency in the real estate market for ultra-expensive homes to be pulled downward in price value to approach the selling prices of less expensive homes in the area. I once knew a retired couple who had spent the majority of their years in a particular area of a large city. They were especially enamored of the area, because it had been the scene of their childhood play and the area where they had courted and married. With their nostalgic memories in mind, they built a rather expensive home in the area, failing to heed the fact that the area was in a state of economic decline. There were a number of low income homes in the area, and there were no homes that even approached the quality or value of the one that they had built. Within five years, the area became saturated with undesirable buildings. It was exceedingly difficult for the couple to find a buyer for their property. When they did find a buyer, the selling price of the property had to be greatly reduced. They suffered a small loss on the sale, but a house similar to theirs in a more desirable part of the city was sold about the same time and returned a profit to the owners of more than sixty percent of original cost.

Nearby low-income housing is another deterrent to the selection of a home in the middle income bracket. For this reason, developers some-times buy the property on either side of their proposed development to preclude the development being surrounded by low income property. The experience of the couple cited above is equally applicable here,for the area where they built not only had houses in a lower price range; the area also had a low income housing area just two blocks down the street.

Concentration of population is an objectionable feature to those in the middle and upper-middle income bracket. For this reason. one should avoid an area where there are large numbers of town houses and condo-miniums, for they translate into high population densities, thereby turning away buyers looking for a little privacy in their lives.

Nearness to commercial or industrial property is another important consideration, because the marketability of property frequently diminishes in almost direct proportion to its nearness to commercial or industrial property.

The location of property with respect to schools is one of the questions most frequently asked by home buyers moving into a community. It follows that a lot should be selected only after the school question is considered, not only the distance to schools, but also the overall standing of the schools within the general area.

Responsible developers invariably incorporate certain restrictive covenants into the contracts for selling their lots. Restrictive covenants can work both ways. They should be carefully examined before one makes a decision to buy a lot in a particular development. It may be advantageous to the buyer to have assurance that there will be no trailers parked in his area, or that there will be no sub-standard homes built on the lots on either side of him, but on the other hand, the restrictive covenant may be overzealous in specifying the precise colors that he may paint his house, or the covenants may prohibit his building a workshop or garage that he may need in the future.

A lot should be in the right area of the city. An out-of-towner does not always know what part of the city is considered the most or least desirable. A friend of mine once bought a lot located on a mountain side overlooking a small city. The view was beautiful in every direction, and was surpassed only by the magnificent flowering vegetation that covered the entire area most of the year. After buying the lot, my friend discovered that it was on the wrong side of the city and was located just off the road to the city dump. It took him fifteen years to sell the lot.

An additional important consideration as far as location is concerned is accessibility the year round. Steep gradients may not be negotiable during inclement weather. Nearby streams may be subject to flooding during heavy rains. One must remember that a home is really not a home if it cannot be used regularly and dependably.

Trees are vitally important to a residential area. The buyer of a lot should concern himself not only with the abundance of trees, but also with the types. Deciduous trees invariably command higher respect than most types of conifers. We have often seen large developments in treeless areas, and we have all noticed the barren look of the area with roofs shining in the noon day sun, with not a single shade tree to break the droning monotony of the landscape.

The prevailing tax rate in an area is worthy of careful investigation, for the differences in taxes among competing areas of a locality can sometimes be significant.

The topography of the land is also important, for it may dictate the types of housing that can be built as well as the density. It may also have a bearing on the cost of construction of the houses, as well as streets and recreation areas.

There are certain danger signals for which one should be alert when looking for a lot. Among these are exhorbitant realtor commissions that are often disguised as other realty fees; developer post development responsibilities not being spelled out in the contract; and soil that will not accommodate septic tanks, or provide adequate drainage.

One should also beware of a lot size or configuration that will not accommodate the house plans he has in mind; deep well water production that is inadequate for regular and emergency water needs; and utility lines that are not immediately available, or cannot be made available without prohibitive expense to the lot owner. Frequently, there is no provision in the purchase contract for utility lines to be buried underground within a development to enhance the natural beauty of the development; and, last but not least is the admonition to beware of filled areas. Many of the foregoing are self evident; however, a few are worthy of additional comment.

Many developers sell their lots only to builders,who in turn include the price that they paid for the lot in the total price quoted for the building; however, the developer tacks on six percent of the total cost of the house and lot when the closing transaction takes place. It is amazing how many unsuspecting people willingly pay the additional expense with no questions asked. A buyer should make certain that he is billed for the lot only once, and should make it clearly known in advance the exact dollar figure that he will pay for the lot. Paying a six percent commission to the developer's representative is an added cost that the client should avoid.

Often, the developer makes rosy promises about his intent to beautify the area, only to walk away from the development and forget all his promises after his last lot sale is made. To avoid this difficulty, the buyer should make certain that the promises are included in the contract of sale for the lot, even then, the situation has a certain amount of risk, for if the developer should go bankrupt, the lot buyers are left holding the bag. Lot buyers have as much right to demand that money be placed in escrow to guarantee certain performances as do the developers.

Many buyers purchase lots in the belief that any type of structure can be placed on any lot. This is a bad assumption. Many lots preclude the incorporation of a basement into the house. Restrictive covenants often prescribe the distance that the completed structure must be from the street, or from the adjacent lot. Another frequently found restriction is an easement crossing the property at some point, thereby precluding any part of the completed structure being placed on or near that easement.

If a lot is located within a city, water pressure problems are generally minimal, because water is drawn from the city water system;

however, structures outside cities that depend upon deep well pumps frequently experience water supply problems throughout their useful lives, because the underground water supply system is inadequate. To avoid this problem, one should inquire of potential neighbors in the development or the general area as to the number of gallons per minute flow that they achieved when their deep well pump was drilled. Even this method is not certain, for it is not uncommon for adjacent properties to vary greatly. My own deep well pump came in at thirty gallons per minute, while my next door neighbor has a flow of only three gallons per minute. However, if you find that every house in a development has low water flow capacity, you may be reasonably certain that the well that you would drill would be no exception.

In general, a buyer should avoid building on a filled area. An often practiced ruse of greedy developers is to buy practically worthless land that is swampy, of poor soil quality, or in deep ravines, and modify it by filling it with every type of debris that they can come by, including trash and rotting vegetation. After they have filled the area to the level of surrounding land without adequate compaction, they sell the lots to builders and unsuspecting buyers who proceed to build a home on the land. After only a year or two, the decaying vegetation causes the earth to settle, thereby causing huge unsightly cracks to develop throughout the structure. When this happens, it is frequently too late to rectify the error; the damage has been done.

Few people consider the direction that the house will face once completed, yet this is an important consideration. Ideally, a house should be so situated that the deck is shaded in the afternoon. Otherwise, much of the enjoyment of outside cooking or late afternoon parties will be denied.

When the time comes to think of signing a contract to buy the lot, and the buyer has duly prepared himself by performing the necessary investigation, one important consideration remains. That is for the buyer to negotiate. Don't pay the asking price. It is not uncommon for developers to ask as much as fifty percent more for a lot than what they actually expect to receive. An in-hand cash offer is generally hard for the developer to refuse. He is in business to make money, and it is frequently in his best interest to sell out as soon as possible, rather than be stymied by waiting for a higher return. When confronted with a lower offer, he may tell the buyer that he has another desirable lot that he will sell for that price. In such a case, the buyer should stand his ground, with a positive answer that he is making his offer on the lot of his choice, not some other lot. It is surprising how often the offer will be accepted immediately. Situations will vary city by city and community by community, but a good starting point in the negotiation process is about seventy percent of the developer's asking price. Be prepared to walk away if he declines your first offer, but let him know that you are interested by leaving your telephone number and telling him that your purchasing the lot is simply a matter of your reaching a fair agreement on the price. Chances are that you will receive a telephone call, either accepting your bid, or making you a reasonable counter offer.

THE BUILDER: The selection of a builder requires far more research than that going into the selection of a lot. Choosing the right builder can make the difference between a dream home and a nightmare. The search for a reputable one must be thorough in every respect. In every area there are local homebuilders associations. One can start by calling them and asking for a list of builders, but this should be only the first step. One can ask friends about builders with whom they have dealt, but, again, one should not hastily accept what friends have to say. Even those people who have had bad experiences are sometimes reluctant to admit it. Nevertheless, builders recommended by friends may be included on the list of possible candidates provided consideration is given to other sources of recommendation.

One of the most valuable sources of information is the scouting of the area to find buildings that are under construction, or have recently been completed. A simple drive-by will eliminate the types and styles of buildings in which the buyer is not interested. The buyer wants to know what builders have constructed the type of house that the buyer wants built. When the buyer sees a house that he likes, he should arrange to examine it, by contacting the builder, the listed realty firm, or even the owner. In viewing the home, one should look for quality of construction, builder attention to detail, the trim work, the parallel alignment of the tops of windows and doors with ceilings, the parallel alignment of sides of windows, doors, and closets with the corners of the rooms, and the swing of the doors to see if they are hinged properly, swing freely, and maintain their position when left partially open. If the owner is available, inquiry should be made as to owner satisfaction with the builder; whether he was satisfied with the quality of material and workmanship; what faults he discovered, even minor ones; whether the house was completed on schedule; and the final cost compared with the estimate that he was given.

After you have identified several potential candidates check with the local Better Business Bureau or Consumer Protection Council to see how many and what types of complaints may have been lodged against the contractors that you are considering. You may also check with the county or city building inspector to see if the building inspector has encountered difficulty in making any of your candidates adhere to the local code. Check customers that the contractor may have provided you as references. Pluses and minuses gleaned during inspections should be carefully noted by the name of each builder. After enough information has been gleaned to narrow your choices to only a few builders, the buyer would be well advised to visit the office of the county clerk of court to ascertain whether there are court judgments or liens recorded against those builders.

It is vitally important to understand that once a reputable builder is located, it is far better to select him, even at a ten to fifteen percent higher price than to select one of the novices of the trade who quotes a cut-rate building fee. When a builder says that he has a building warranty, ask for it in writing and examine it closely, but don't be misled by its contents. Historically, home owners have had rather poor luck in getting the builder to return to the scene without an imminent threat of court action, and remember that most contractor defaults are considered civil, not criminal. A threat of legal action against a builder who has all his assets in his wife's name is no threat at all.

After you have narrowed your list to three or four contractors with whom you would feel comfortable, ask each of them for bids. Make certain that each has a complete set of specifications and a complete material list where the materials are precisely defined. When all bids are in, examine each bid separately and thoroughly. Don't automatically select the contractor with the lowest bid. In commercial construction, many long time entrepreneurs automatically reject the lowest bid, saying that almost invariably, the low bidder has failed to consider some important aspect of the work or he plans to do the job with less than top-notch personnel. If you eliminate the low bidder, you will then have to discriminate among the other three or four. Look to see if there is one that is at least ten to twelve percent higher than the others. If this proves to be the case, eliminate him also.

After narrowing the field to only about two general contractors, ask each a number of important questions such as how long he has been in the business, his overall record of achievement, whether he is bonded and licensed for the type of work involved, the type of insurance that he carries, whether his workers are covered by workman's compensation, and any other questions which you think might help you decide from among the competing contractors. Before finally deciding however, you should take a look at some of the work that each candidate has done and talk to any of his former clients that you have not already contacted. By then you should have your own strong feeling as to the contractor with whom you would feel the most comfortable. Saving money is important, but it is always a matter of degree, if the final bids are only $2000 to $3000 apart, money should not be the prime consideration. It should be overridden by your own feeling of how much you can trust one or the other and how comfortable you feel in discussing various aspects of the job with each of them. Remember that the signed contract should specify precisely what is to be done by the general contractor, and when it is to be done. It should specify the type and grade of heat pumps, furnaces, hot water tanks, kitchen appliances, plumbing fixtures, and all options that the client wants for the house. If the client is not sufficiently versed in describing the items that he wants, he should go to the vendors of the items, identify the manufacturer and model number of his choice, and then incorporate into the contract statements that the general contractor is to provide the items that he has specified or items of comparable cost and efficiency. If the client wants the exact item, he should eliminate the word "comparable" and specify the precise model that he has selected.

Once you have thoroughly evaluated the top builders on your list and have made a tentative selection, you should have a credit check performed on that particular individual. If the credit check comes back with derogatory information, you can then consider the second builder on your list, whom you would then make the subject of a credit check.

It is impossible to over-emphasize the importance of selecting the right builder. The entire process of contracting for the building of a home centers upon the success or failure of following the proper procedures in dealing with potential recipients of your contract. Many people contract for the building of a home, by observing the necessary precautions in selecting a building lot, but, in their haste to get the building process started, or to save a few thousand dollars, they select builders who turn their quest for a home into a nightmare. This book is designed to set down guidelines that will prevent future clients from experiencing those nightmares.

If my wife and I had followed the above advice, it would have saved us thousands of dollars and two years of continuous heartbreak. As will be revealed in detail later in the story, we selected a builder who quoted a cut-rate fee and was recommended by a homeowner for whom the builder had recently built a house. We examined two houses that the builder had under construction and proceeded blindly and ill-advisedly into the trap which he had set.

The rest of this chapter deals with the ill-fated experiences of my wife and me in selecting a contractor. In our haste to get the building process started, our selection of a builder created problems which proved to be almost disasterous.

We found our ideal lot on the outskirts of the city. The front yard had a huge old oak tree, estimated to be about four hundred years old. We purchased the lot, then we figured we were ready to hire a contractor. Our chief obstacle was that we really didn't know how or where to start looking. The entire area was in a rapidly expanding phase and it seemed that most of the reputable builders already had more business than they could address for several months. The first builder that we contacted quoted a construction fee that was a full $35,000 in excess of what we found had been paid other contractors for building similar type houses. Another builder came to our house, reviewed the plans and promised to return with a bid within a week. We did not see him or the plans again. Locating a competent and dependable builder was proving to be a more formidable task than we had anticipated. Because of our anxiety, we contacted several builders who had not established a reputation in the area. One of these was a young personable builder who displayed a keen interest in the job and promised to provide everything we requested in minimum time with construction to begin within a week. We were on the verge of awarding him a contract when a neighbor informed us that the builder was in deep financial trouble as a result of having over-extended his construction volume at bargain prices. The builder tended to corroborate this report when, on his next visit, he asked to borrow $3000 against his future contract so he could pay off the loan on his truck, which he said the loan company was about to repossess.

Our continuing search led us to a site where a new building was under construction within the development which we lived. The builder, a stout and totally bald little man greeted us enthusiastically and introduced himself as Beryl Jaker. He explained that at that time, he was not only constructing the house in which we were standing, but was building another house less than one mile away for a research scientist. He invited us to visit the other building, stating that it was about seventy-five percent complete. He said that he would build the house which we had in mind at a considerable savings, compared to fees being quoted by other builders.

We drove immediately to his other building site. The building sat approximately 150 yards off the main road in a new development consisting of new $250,000 to $300,000 homes. It was accessible by a winding graveled road that had been cut through a pine forest to a large clearing at the end of a cul-de-sac. It proved to be about a 4000 square foot contemporary building with a number of innovations not commonly seen in the area, displaying a flair for the dramatic in almost every room. To untrained eyes, the construction seemed to be quality work and it appeared to have a certain style that was not readily identifiable by either of us. Only a few months later we were to realize just how naive we had been that day. We were to become painfully aware of things that we should have examined and questions that we should have asked of a more knowledgeable person within the building profession.

The research scientist for whom Mr. Jaker was building the house was a Mr. March. He informed us that Mr. Jaker had been highly reliable in every respect; that Mr. Jaker had several years of building experience, but that because of a series of misfortunes in Battle Creek, Michigan, Mr. Jaker had been forced into bankruptcy, and had moved to this location to start all over again. He said that he had found Mr. Jaker to be impeccably honest, and that even though Mr. Jaker had not attended architectural or engineering school, he had developed the talent to do his own design work. He suggested that Mr. Jaker would be an excellent selection.

Without investigating Mr. Jaker any further, we called him the next day and asked him to visit our home to discuss the details of a contract. The fact that he had declared bankruptcy in Michigan should have served as a red light to us, but we were so anxious to get started on our home and we were so favorably influenced by the laudatory remarks of Mr. March that we stumbled straight forward into the trap that was to cause us so many hours of pain and hardship during the next two years. Of course we were influenced by the feeling that we could employ Mr. Jaker for several thousand dollars less than we could other builders. For some reason, I felt a certain distrust of Mr. Jaker. I told my wife, "I can't identify the feeling. He is not an educated man but he is smooth of tongue and seems to feel that no obstacle is too great for him to hurdle with the greatest of ease." I suggested that it wasn't too late to consider the young man who had wanted to borrow the money to pay off his truck; that I believed we could employ him for approximately the same fee as Mr. Jaker; and I suggested that we might be able to work around his financial difficulties by my handling all the funds for construction of the building, including the acquiring of the construction loan in my name. My wife replied that we knew that the young man was in financial trouble and she would feel more secure with Mr. Jaker; that he was an older man and more experienced.

We were to learn later just how experienced Mr. Jaker was not in the building business, but in the misuse and misapplication of funds associated with building. In fact we were to learn that Mr. Jaker was a bonafide artist in bending the civil laws of our jurisprudence system to fit his needs, and in such a manner as to render him immune from criminal prosecution, however heinous his crimes. Later investigation would reveal that he had been a policeman in Battle Creek, Michigan for a number of years; that he had been expelled or released from the force; that he had become involved in the building trade immediately thereafter; and that he had been the object of several law suits brought by people with whom he had contracted to build homes. This infamous activity had ultimately led to his declaration of bankruptcy and his move to another state to ply his trade in an area where he was not so well known.

When Mr. Jaker reviewed our plans, he assured us that there was nothing that would pose any problem whatever; that he had already identified a number of facets in the design upon which he could improve; that he could start work the following week; and that the house would cost us no more than $206,000 including his construction fee. He said that his entire fee for building the house would be only $18,000. He suggested a COST PLUS FIXED FEE contract, which of course meant that Becky and I were to pay all costs including those associated with unsatisfactory work, although we weren't cognizant at that moment of the entire cost plus fixed fee procedure.

Our contractor returned two nights later with a crude contract which he had prepared himself. It reiterated his earlier statements of a $206,000 cost including his $18,000 fee, and contained the stipulation that the house was to be completed seven months after completion of the foundation. What we didn't realize at the time was how long he intended to procrastinate in completing the foundation, or that there were no penalties involved even if he did not complete the building in any appointed time. We were aware however that should completion of construction require several months more than stated, Becky and I would be losing some $1,800 per month in interest. So, the following day, I asked David Hoffer, our attorney to review the contract, to advise us of any pitfalls, and to reword the contract so it would be in the proper legal format. He did nothing but reword the contract. He saw no pitfalls whatever. Mr. Jaker, the former policeman and dishonest builder would be pleased with the legal document that my attorney had prepared. There were several elements of the contract about which the attorney should have been wary; foremost among them were the stipulations that Becky and I were to provide Mr. Jaker with $30,000 cash to finance building expenses while he was obtaining the necessary construction loan, and the stipulation that we were to deed the lot to Mr. Jaker and his wife at the start of construction, so that the entire transaction could be processed as a builder to buyer operation when the building was completed. For the $30,000, Mr. Jaker and wife were to give us a promissory note. After David Hoffer provided us the requested legal document, I called Mr. Jaker and asked him to visit our home to sign the contract. Mr. Jaker assured us that work on the building was to begin the following week. He promised to consult with me on all bids that he obtained that were in excess of one hundred dollars. He also promised to provide us with receipts every three or four days so we could be informed of the costs of labor and materials every step of the way. Succeeding events proved that he had no intention of keeping any of these promises.

Before continuing further with the details of the adversity suffered by my wife and me, we must return here to the informative aspects associated with contracting for the building of a home. The next chapter discusses the various types of contracts which may be employed to build the home and discusses some of the most prominent advantages and disadvantages of each type of contract.

THREE: FIXED PRICE OR COST-PLUS-FIXED-FEE

Before even thinking of signing a contract, it is necessary for the person interested in having a house built to be familiar with a few of the most basic terms. The person who contracts with the owner is known as a general contractor. They may call themselves builders, but legally, they are known as general contractors. The army of craftsmen and skilled laborers whom the general contractor employs to perform specific tasks during the construction process are known as subcontractors. Legally, subcontractors are mechanics. If they are not paid for their services during the construction of a house, they may place what is known as a "mechanics lien" against the house. The placing of such a lien generally means that closing on the house, or transferring the property from one person to another cannot take place until the mechanics lien is paid.

Vendors who supply material for the construction of a house are known as "materialmen". Among these are people who supply concrete, lumber, insulation, drywall, carpeting, and other materials used in house construction.

As previously noted, there are two principal types of contracts that may be consummated between a builder and client. They are the fixed price contract and the cost-plus-fixed-fee contract.

The fixed price contract is one whereby the client agrees to pay the builder a set price to build the house specified in the plans. If the client has followed the correct pre-contract procedures, he has acquired a set of detailed specifications from the architect or house designer. The client then assures that the contract stipulates that the structure is to be built strictly in accordance with the specifications which he has provided. Theoretically this should assure the construction of a quality building. It will, unless a builder has been selected who is intent upon and adroit at omitting the installation of costly items stipulated in the contract, or at the very least, substituting cheap material where higher grade material is specified. Building costs ascend rapidly during the construction progress; hence, the builder profits by omissions and substitutions at each step of the process. It is important that a client be aware of the possible employment of some of these omission and substitution techniques, so he can identify them if they occur and exercise a degree of quality control.

Things which the client can do to assure quality control during the construction process:

- Check the plans to assure the installation of adequate concrete footings in the proper size and number to support the super structure.

- Make certain that the general contractor compacts the soil where soil density is not adequate for long term support.

- Inspect each shipment of lumber to prevent use of rejected, weather damaged, poor grade, or previously used lumber for framing and subflooring.

- Examine the insulation sheeting to insure that it conforms with specifications.

- Insist that plywood be applied as bracing for all corners of the structure.

- Check to insure that vertical members are not bowed,split or out of plumb.

- See that holes are drilled in stringers and studs for the installation of electrical wiring, don't permit the construction crew to staple the wiring to the bottoms of floor joists and the outside surface of wall studs.

- Look at all molding to see if it conforms with specifications.

- Don't permit the installation of finger molding unless it is specified.

- Don't permit substitution of masonite doors for more expensive and more efficient wooden doors.

- Check the grade and cost of floor tile, hardwood flooring, wall paper, paint, carpeting and cabinetry to assure prime quality.

- Perform your own inspection to insure that insulation is installed in corners and other areas of the house where accessibility is difficult.

- Insist upon full allowance for light fixtures and carpeting when time comes for you to make your selections.

- Insure that metal trunks are installed for the primary areas of the heating and cooling system, and that the house includes a heat pump, furnace, and water heater as prescribed by the specifications.

- Make certain that the roofing material conforms with the specifications, and that a lighter and cheaper grade has not been substituted.

- Check the grades and qualities of windows. High quality windows have two layers of safety glass with air pockets between them to prevent loss of heat. Cheap windows do not provide this protection. Also, the proper installation of windows necessitates their being encompassed in plastic sheeting around all edges of the windows where they fit into the wall openings. This action prevents the accumulation of moisture around window edges. If the remedial action is not done correctly, it will cause the edges of the window to decay. Many contractors eliminate this vital step, knowing that the buyer will never know the difference, until three or four years later when he discovers excessive window rot.

- See that the builder treats outside doors with quality primer paint or polyurethane. Sunshine and rain can be destructive to doors facing the outside.It is essential that such doors be given from one to two coats of primer paint or polyurethane to prevent the absorption of moisture. Without such protection, the doors will come apart in a matter of about two years, thereby necessitating replacement at a cost of about $200.00 per door.

- See that the contractor pre-treats stained woodwork with a proper conditioner before applying the stain. If he doesn't apply such conditioner, the stain permeates the soft wood unevenly and results in an ugly mottled appearance of the finished product. The staining of soft wood not only requires a high degree of skill; it is extremely time consuming. It is amazing how many $300,000 to $500,000 homes have mottled wood work.

A few years ago, while I was living in Northern Virginia, I noticed that during winter weather, my family room remained extremely cold and damp, regardless of how high I elevated the thermostat.

Finally, in desperation, I went into the unheated garage which was adjacent to the family room, and cut out a portion of the wall board to see if the wall had been insulated at the time the building was constructed. I found that the wall did not contain a shred of insulation. This discovery led me to examine the area underneath the house to determine whether the flooring had insulation underneath. I found that it did not. In spite of the fact that the city building code specified that all flooring and all exterior walls were required to be insulated, the builder,a nationally renowned corporation, had failed to insulate. Having experienced the shocking discovery in Northern Virginia, I observed the practices of a builder who was building a house next door to mine in North Carolina. I noticed that he was short cutting his clients in the insulation of the building. I advised the clients of my observation, only to have them respond by silence. They obviously thought I was a busy body who should be minding my own business. After they occupied the premises,we learned that their electricity bill tended to average about $365.00 per month, while mine averaged about $135.00 per month. Both houses had approximately the same square footage.

An unscrupulous builder can pocket huge profits during the construction process. It is the business of his client to assure that the profits are those earned by conforming with the specifications and providing the highest quality product.

In a later chapter, suggestions will be provided regarding possible ways for the client to engage professional help to assure quality control of the general contractor's work. The fixed price contract is the type most often used by those contracting for the building of a home, because it is the type with the least risk. It does not have the potential to damage the client as severely as the cost-plus-fixed-fee contract, which has the capacity to destroy the life's savings of the client.

The cost-plus-fixed-fee contract is a negotiated agreement whereby the client agrees to pay all costs associated with the building, including but not limited to all material, all labor, all services, and all interest on the construction loan. In addition, the client agrees to pay the builder a set fee for his supervision and expertise. This is a contract which should be avoided by everybody, except under the most unusual circumstances. In some cases, the client agrees to pay the general contractor a percentage of the total costs of the building. For instance, if a client agrees to pay the builder twenty percent of the total costs of a building and it turns out that total costs are $200,000; then the client will owe the builder $40,000 for his services, thereby making the final cost of the home $240,000. This contract has the actual effect of providing the builder a signed check in unlimited amounts. It means that the client is agreeing to pay whatever price the builder agrees to pay for whatever material or service for the house. It also means that the client is to pay for all mistakes that the builder might make, either through negligence or lack of knowledge, for when an error is made, it must be rectified, and all the labor and material to rectify it, is at the client's expense. The biggest and most dangerous aspect of the cost-plus-fixed-fee contract is the fact that it gives the builder carte-blanche permission to borrow any amount of money that he can borrow against the client's building, thereby subjecting the building to the imposition of potential liens that may exceed the value of the building. In a case where the client agrees to pay the builder a percentage of total costs, it is an invitation to the builder to pay top prices for all material and services going into the building so that he may increase his profit margin.

In other cases, builders execute what, they call a "roll", an act whereby they borrow maximum funds against one building, but apply them to the next building on which they have started construction. When settlement time comes, the client may find that the building has cost him fifty to seventy-five percent more than the contract originally specified. The client will then have to decide whether to pay the amount of the liens to keep the lien holders from foreclosing on the building, or to sacrifice whatever amount of money he has already invested in the building.

The law says that whatever misdeeds the builder commits, he commits no criminal offense; that the offense is only civil. It is a common practice in many localities for one desiring to contract for the building of a house to deed his lot to the builder and his wife, and to advance the builder a sum of money amounting to approximately fifteen percent of the pro- jected cost of the completed building. The builder can then obtain the construction loan by providing the lot as security. The advance of money is supposed to defray start-up costs that accrue while the construction loan is being processed. In return for the advance of money, the builder and his wife sign a promissory note agreeing to pay the client back when the building is completed. This is a custom that is fraught with enormous potential for misuse of funds and outright theft, but hard as it may be to believe, it is common practice. As will be revealed later in this book, clients taking such contracts to lawyers for review are frequently advised to sign the contract.

There are eminent perils that may be inflicted by a self-seeking builder, regardless of whether he is operating under a Fixed Price contract or a Cost-Plus-Fixed-Fee contract. These perils can be avoided by the client doing the necessary work to assure that the contractor he has selected is a reputable one with a history of satisfied clients.

The question arises as to whether it is better for an individual unfamiliar with the dangers of contracting to purchase a ready built house, thereby avoiding the dangers associated with the contracting operation. The answer to the question is that it depends upon the circumstances. Thousands buy ready built homes each year, many of whom are perfectly satisfied with their purchase and would trade for few other homes; however, case histories are replete with thousands more who buy new homes only to discover too late that their house has not been accorded the high quality workmanship promised by the developer or the builder from whom their new home has been purchased. The dangers associated with buying ready built homes are outlined in the next chapter.

FOUR: PURCHASING A NEW HOUSE

Purchasing a ready built house is by far the mode most frequently chosen by people wanting a new home. In many respects, this mode is comparable to engaging a contractor to build under a fixed price contract. Many of the dangers are the same. There are a number of potential perils in buying a ready built home. In the previous chapter, it was pointed out that builders frequently obscure structural defects and other construction flaws by covering them with wallboard or paint. This continues to be true in the case of some ready built homes. A builder who is so inclined has even greater opportunity to increase profits by decreasing quality when he builds a home for sale on the general market.

Ready built homes are frequently identified by builders as "spec houses", because builders speculate that they will find buyers quickly and that they will realize nice profits on the buildings. Originally, the term "spec house" referred to houses that were built with concrete or fixed specifications. Gradually the meaning changed to identify mass produced or rapidly built homes built for investment type speculation. Builders sometimes become overzealous to increase the profit margin on the finished product. All the omission and substitution ploys identified in Chapter 3 may be used. They sometimes install inferior insulation. incorporate poor grade or used plywood, install cheap roofing, use mis-tinted or watered down paint, purchase poor grades of lumber, masonite doors, cheap hardware, inferior windows, off brand heating and air conditioning equipment, and even the cheapest grade of electrical junction boxes. They save on services by employing subcontractors who are not fully skilled in their trades. Inasmuch as there is no client to inspect their work while the house is under construction, they have only the city or county building inspectors to appease, a task that is relatively simple when performed by seasoned builders.

When the spec house is completed, it is turned over to a real estate agency which agrees to sell the house for a six percent commission, or less. Regardless of the language in the builder-realtor sales contract, there is generally a full six percent commission tacked on to the final sales price, which is a fee that the buyer does not have to pay when he contracts for the building of his own home. There are a number of things for the house buyer to remember when he starts to buy a new ready built house. Many of these things are identical to those for which he must be alert if he employs a builder under a fixed price contract. The difference is that construction flaws under the fixed price contract can often be identified at the point that the error occurs, whereas, in the purchase of a new building, most of the flaws may be undetectable because they are covered by wallboard, siding or paint. However, there are a number of things for which a potential buyer should look when contemplating the purchase of a ready built house. By closely examining these features, one can still do a pretty good job of determining whether a house has been built with high quality workmanship and materials. The following list provides guidelines not only for inspecting a ready built house; the same checks can frequently be used to ascertain quality of work performed by subcontractors or general contractors:

1. Ceilings: Check to see if the blown-on paint is uniform; i.e. that it is not thicker in one area than in another, or that there are no spots where the ceiling is almost bare. Make certain that there are no holes that show where the drywall has been nailed. Also look for small bumps on the ceiling that indicate that a nail is about to pop through the wallboard.

2. Walls: Make the same nail hole checks that you made for the ceiling. In addition, the entire surface should be smooth. There should be no evidence of where the wall board tape was applied to seal the cracks between sheets of wallboard. Look at all the electrical receptacles and switches to make sure that the electrical plate completely covers the hole that was cut into the wall for installation of the junction boxes.

3. Windows and doors: Check to see that all window corners are square, that is an exact ninety degree measurement. If they aren't square, you may have trouble opening and closing them without considerable difficulty. Look at the doors to see if they swing freely on their hinges and if they latch and unlatch easily. Check to see that there is no excessive space between any edge of the door and the facing. Such a space can permit cold or hot air to enter or escape and thereby create high energy costs. Check all doors to see if they are warped or have hairline cracks that will enlarge as time passes. Check to see that all outside doors have weather-stripping all the way around the door. Close the door to see if there is a tight fit with the weather-stripping.

4. **The Kitchen:** This is the area that most women will examine if you ever put the house on the market. It is important to understand that construction crews are sometimes careless. They frequently scratch, dent, tear, or break appliances and/or material during the construction process. So, look closely at every item to see if it has been damaged in any way. Turn on each appliance and see if it works. Look to see if any of the ceramic tiles have been broken. Look at all the joints to see if they have been filled with grout and the grout has been neatly and uniformly applied. If the kitchen floor is vinyl or asphalt tile, see that the joints run in straight lines and that squares match adjacent squares. Make certain that shoe molding has been installed where the floor meets the walls and cabinets and that it has been caulked to give a smooth neat appearance. If the floor is covered with vinyl sheeting, see that the seams are well obscured, that they are hardly visible, and that all the material runs in the same direction. See that every seam is well cemented. On wood floors, look for scratches or gouges where the appliances may have been moved around. See that the finish on the floor has been applied uniformly. There should be no chipped or broken tiles on the kitchen counters. Check to see that there are no loose edges. Thoroughly check all kitchen appliances to see if they work properly and to see if they have been damaged in any way that will detract from the beauty of your kitchen. Examine the front of the dishwasher for scratches. See that the dishwasher has been securely installed in its cabinet and that the door operates freely.

Look for dents in the oven and range and see that all the inside parts are present. Examine glass panels to make sure that they have not been chipped or cracked. Look in the hood above the range to make sure that it has been vented to the outside. See if the opening for the vent may have been cut too large thereby providing an opening for cold air to enter the house. Check all other appliances for damage and proper functioning, making certain that all parts are present. Carefully inspect all the cabinets for marred surfaces, dents,cracks, and missing door and drawer handles.

Check also to see that the cabinet doors open and close properly. Pull open every drawer and door to see if they open and close properly.

5. **Bathrooms and Laundry Rooms:** First, check to see that all faucets work properly; that they don't drip when turned off; that faucet handles are installed properly; and that you can obtain a proper mix of cold and hot water without difficulty. Hot water should come from the left spout and cold water from the right. If there is only one spout, see that the

handle on the left provides hot water and the one on the right gives cold water. Check the lavatories and bathtubs for scratches and dents. Check counter tops in the bath rooms and laundry rooms the same way that you checked those in the kitchen. Examine lights and lighting fixtures for broken or chipped panels. See that all lights respond properly to the switches on the walls. Shower heads sometimes leak where they are joined to the supply pipe. Examine the area where they are attached to see that there are no leaks. Examine the shower drain to be certain that it is sealed where the water can't leak to the area underneath the shower. Examine toilet seats for breaks and damages and to see that they raise and lower properly. Examine medicine cabinets for brakage of glass and for scratches and dents. Check to insure that the cabinets have all shelves that they are supposed to have. Look at all electrical receptacles to make certain that the holes have not been cut too large to be covered by the electrical plate. In the laundry room, perform the same checks that you did in the bathrooms. The walls and ceilings should have uniform painting, and lighting should be adequate. Shoe molding around the base should be properly installed. The laundry room should be equipped with a 220 volt receptacle for your dryer and at least one 110 volt receptacle for your appliances and washing machine. It should also have a drain pipe for the washing machine and an exhaust vent for the dryer. More expensive homes will have laundry tubs installed.You should check all these items to see if they function properly.

6. The Electrical System: Ensure that all switches operate the appliance or light fixture that they are intended to operate.Insure that all fixtures have required accompaniments such as bulbs, shades,and globes. See that all cover plates have been installed.See if all three and four way switches operate properly. Most electrical systems are equipped with a ground fault interrupter(GFI). You can test a GFI by pushing the test button. This action should shut off the power to the entire circuit. You can then screw a light or plug in a tester to each of the outlets on that circuit and see if they have been turned off. If they have, the wiring has been done correctly.

7. The Attic, Basement, and Garage: If the garage has an electric door opener, see that the remote control works satisfactorily and that the door goes up and down smoothly. If no electric door opener is installed, check to see if an outlet is provided for future installation. See that the floor of the garage is at least four inches lower than the adjacent floor level. This difference in levels is to prevent water from entering the house if it should get into the garage. The garage floor should be smooth and screeded in such a manner as to cause water to flow toward the garage doors instead of toward the house, or instead of puddling in the middle of the garage. For the attic, the builder or his representative should have provided a certificate stating the R rating of the insulation in the house. These R ratings are described in Chapter 5. You can verify these ratings by entering the attic and checking. You can see the thickness of the ceiling insulation, but the walls will be already sealed. If the entrance to the attic is by a hatch door or disappearing stairs, make certain that the inside portion of the door or stairs is insulated; otherwise, you can expect to lose a lot of heat during the winter months and a lot of cooled air during the summer. If you have a basement, the important consideration is that it be a dry basement. To inspect for dryness, look to see if there are water stains on the concrete block or on the wallboard. Look also for large cracks in the basement walls, or for the formation of mildew around the base of the wallboard. Ask for a guarantee against basement leakage when you sign a contract. In many jurisdictions, basement walls do no have to be insulated if the basement ceiling is insulated to keep heat from escaping downward out of the house or cold air from going from the basement through the ceiling into the living area above. Also, many codes specify that if the basement walls are well insulated, the ceiling or the basement does not have to be insulated. If you make early enough contact with the builder, you may be able to get your choice of the two systems.

8. The Exterior of the House: The outside of the house will generally be of brick, stone, stucco, wood or masonite. If its brick, check the work of the mason to see if mortar lines are neat, level and uniform and not slopped over the bricks which they are joining. Look for cracks in the mortar joints between the bricks.

You do not have to be an expert to ascertain whether a house has good brick work. The quality will reveal itself. If the work is neat and uniform with level lines, it will be readily apparent. If it doesn't have these characteristics, there are grounds for assuming that the rest of the house may have been assembled in a sloppy manner also. If the house is of stucco, look for fairly large cracks that may have developed in the stucco siding. Don't worry about small hairline cracks. If large cracks exist, you should demand that the builder have them repaired. On a house with wood siding, look to see that everything is uniform; that vertical members such as board and batten are installed parallel with a constant distancing between vertical members and between horizontal members. See that the color is consistent throughout. At all points where different materials join, there should be caulking or a good sealant applied. At points where plumbing or electrical lines enter the house there should be a caulking material to lock out the elements from the holes through which the plumbing or electrical lines have entered the house. See that all doors have been properly weatherstripped so that when closed, there is no air entering from any point around the door.

Make sure that ventilation holes in the roof have insect screens over them to prevent birds and insects from entering. If the house has a basement with window wells, make sure that each window well has a base of gravel instead of dirt. The gravel should be sufficiently thick for water accumulation to seep into the gravel and away from the house instead of into the house.

See that the yard is graded in such a manner that water falling from the roof will flow outward from the house, not into it or under it. The earth should also be packed firmly around the sides of the house, but should not be graded too high around the exterior walls. At least six or eight inches of foundation wall should show between ground level and first floor level.

Examine the roof to see that the asphalt or fiberglas shingles are not curled, split, or separated. Check to see if gutters and downspouts have been securely attached to the building and that they exit into drains or onto properly positioned splash blocks. Examine driveways, patios, and sidewalks for cracks or chipped corners. If they are of asphalt, see that there are no depressions (birdbaths) which will collect and hold water and that they are graded so that they divert the flow of water away from the house.

In addition to the above checks, there are a number of other factors that you should consider if you are thinking of buying a new ready built house. After all, the one thing that you are looking for, more than any other is quality workmanship. Next to that, you will be looking for low cost maintenance. The questions below are aimed at these two fundamentals. It is therefore important that you consider each of them not only if you are buying a new home, but also if you are intent upon having one built by a general contractor.

Is the builder one who has been in the trade for a number of years?

Every community has reputable builders who take pride in their profession and consistently produce high quality homes; however, the building profession abounds with novices in the trade who have never built a single house and yet, in one way or another, obtain a contract to build their first house. When they build their first one, the subcontractors have a hey day, charging excessively for their work, and producing low quality results. In one way or another, these results are passed on to the buyer in the total price of the house or in the buyer having to pay for correcting the defaults that may not surface for several months. My next door neighbor was unfortunate enough to buy a $250,000 home erected by a first-time builder. My neighbor has been in the house only one year, and his problems have been manifold. He hardly finishes repairing one item when another building deficiency develops.

Has the builder earned a good reputation in the community where he is building? Can the builder identify a number of well satisfied former clients whom he recommends as references?

A builder's reputation and quality references are probably the most important of all the characteristics one should demand of the builder. The building or buying of a home is one of the most important endeavors a person will attempt in his lifetime. One must therefore approach the endeavor with the greatest caution of his lifetime. If a builder has not established a reputation and can provide no quality references, it is almost a certain phenomenon that the buyer or eventual inhabitant of a house that the builder has constructed will be penalized for the builders lack of knowledge or misdeeds. A few days of investigation by the buyer is certainly worth the effort.

Precisely whom is the real estate broker representing, the builder or the developer, or another entity?

It makes a difference. If the broker is selling for the builder, one may have increased chances of having the builder return for remedial work that may be needed. One also has a chance to check other buildings constructed by that particular builder, and thereby ascertain whether his work has been of high or low quality, or if other buyers have been satisfied with his product. If the broker is representing a developer, one is alerted immediately to the possibility of hidden charges that may be incorporated into the cost of the building, especially the six percent charge identified as the realtor commission in Chapter 2. As previously mentioned, it is often extremely difficult to get a developer to return to the site for needed repairs when he has finished his work in a particular development and has moved to another site or to another city. There is a development in a local area today whereby a lake excavated by a developer did not drain properly and the courts passed to the inhabitants of the community the costs of correcting the faulty drainage, a sum of ten dollars per home owner per month ad infinitum.

Is the seller of the property financially solvent?

If he is not financially solvent, there is always the possibility that there may be liens or encumbrances recorded against the house. One must also consider the fact that should litigation be necessary and the seller is financially insolvent, a court judgment against the seller would be worthless.

How does the selling price of the house compare with others in the area when computed on the basis of dollars per square foot? If the dollars per square foot appears to be higher than other houses of the same general size and quality, what is the reason for the higher cost of this particular house? Are the benefits of added appurtenances great enough to justify the difference in price between this particular house and others?

There is a tendency in the real estate market for brokers to set the same dollar per square foot value on property regardless of the location of the property. In other words, if property in the most desirable area of the city has been selling for $90.00 per square foot, real estate brokers tend to set $90.00 per square foot as the mode for establishing the price on property in far less desirable locations. The buyer must be aware of this anomaly and be guided accordingly. There is also a propensity for a broker to set prices far out of proportion to the value of some of the appurtenances in a house. A swimming pool valued at $4,000 should not enhance the value of a house more than that $4,000. In many cases, it should not increase the value of the house at all. In some cases, it may actually decrease the value of the property.

Is the seller recommending a particular mortgage financing agency, and if so, why is he recommending that particular agency? Is there a possibility of collusion between builder and lender in the form of rebate to the builder?

Often, tacit agreements exist between the realty firm and lending institution, whereby rebates are given to the realty firm for mortgage loans directed to the lending institution. The institution concerned may or may not have terms as good as other institutions in the area. A few years ago in Lexington, Kentucky, I was steered to a particular institution whose mortgage loan officers were so self assured that they were even unpleasant to some of the customers. I discussed a mortgage with them, but before signing the necessary documentation, I decided to check several other lending institutions to ascertain whether I could obtain better terms. I was successful on my first try, and I took a certain amount of pleasure in returning their paper work and informing them that I had simply walked two blocks down the street and obtained better terms.

What written guarantees is the seller willing to provide as part of the sales contract? What is the likelihood that the seller will be available a year in the future and financially able to carry out his guarantees? Are there any liens or encumbrances in effect or pending against the building? If so, is the title insurance worded in such a manner as to guarantee that the buyer will not be held responsible for any of the liens or encumbrances?

The answers to these questions are essentially self explanatory. The most important consideration here is the title insurance aspect. Every day, millions of dollars are paid at numerous real estate closings in the United States under the assumption that for the fee paid, the title insurance company will guarantee title to the building. However, there have been instances wherein the title companies have pointed to fine print to deny their obligation. For this reason, the buyer needs to know as much as possible about the seller or builder, or both. It is a great deal of trouble for a buyer to visit the office of the clerk of court to ascertain whether there are liens or encumbrances against a building, but in a lot of cases, the trip will prove worthwhile. One may ask the question, "Isn't this a service for which I am paying an attorney?" The answer is, of course, that it is one of the things for which you pay an attorney, provided you have an attorney. If you don't have one, it is easy to do yourself and you save the attorney fee.

Can the owner provide all the manufacturer warranties on appliances and hardware installed in the building?
Practically all appliances have warranties, but all too often, they are lost during the construction process. Characteristically, a worker installing a new appliance in a home under construction pulls all the documentation out of a box when he starts to install the appliance. As he has installed items of that nature several hundred times, he doesn't need the directions, so he throws all the paper aside to be swept out of the house with the other trash and construction debris. For this reason, it is important for the buyer of the house to demand that all warranties to all appliances in the house be handed over at time of settlement.

How many years is the roofing guaranteed, and under what terms?
There are at least a hundred different brands and types of roofing, varying in costs from as little as seven or eight dollars per square (100 square feet) to as much as fifty to sixty dollars per square. Some may be guaranteed for five years; other roofing for thirty years. It is important for the buyer to know the difference. You may be certain that the builder of a spec house is not going to spend a lot of money to pay for high quality roofing when he knows that the buyer won't recognize the difference.

Does your investigation indicate that the builder is one that would willingly return to the house to perform remedial work if something goes wrong after closing?
As stated in Chapter 2, it is a rare occasion indeed when the occupant of a new home is able to get the builder to return to the house to effect remedial repairs. There are a few who will honor their obligation if the requirement stems from their own work. If the requirement to return for such work is entered into the sales contract, the buyer will have more persuasive capability than he would have without such an entry. At any rate, the effort is worthwhile. A buyer should stipulate in the purchase offer that it is conditioned upon the builder returning if a situation

develops that may be attributable to his workmanship.

Does the house come with a Home Owners Warranty? If so, does the fine print limit its application?

Today, it is fairly common practice for developers to include a ten year Home Owners Warranty with the sale of a house. The buyer should always inquire as to the warranty and exert maximum effort to become familiar with its content before settling on the house. If there is something omitted that the buyer feels should be included, it should be made the subject of negotiation before closing.

What is the quantity and condition of shrubbery and vegetation around the premises? Is the seller prepared to guarantee the shrubbery and vegetation for a reasonable period after selling the house? Frequently, builders employ subcontractors to landscape the houses for sale. Shrubbery and other vegetation set into the ground upon completion of building generally require care for a period afterwards. Unfortunately, post planting care is not part of the itinerary of the builder or the subcontractor. If the house is not sold immediately after construction, the shrubbery may die or become diseased from lack of care and nourishment. The buyer should examine the condition of the shrubbery and other vegetation prior to signing the sales contract and should stipulate the requirement that the shrubbery and vegetation be guaranteed for a specific period of time.

The answers to the above questions are vital to the would-be new home buyer. A negative answer on any one of them could not only result in tremendous outlay of financial resources, it could deny the homeowner much of the joy associated with owning a new home.

Make no mistake about it, if the builder is not committed to carrying out his after-sale obligations to a buyer, a team of wild horses pulling in tandem will never get him to return to the site. During our married life, my wife and I have purchased eight newly built homes in various parts of the United States, and I can't think of a single instance in which we were successful in getting the builder to return for remedial work after the closing transaction had taken place. In one instance, we purchased a home in Fairfax, Virginia, a suburban Washington, D.C. community. The development consisted of approximately three hundred new homes in the middle income class. At the time of sale, none of the streets had been completed. They had merely been scraped out of the raw earth by a dozer and road grader. The developer assured each buyer that he would start pouring the concrete the following week, yet months passed and the scraped areas remained seas of mud or dust, depending on the turns in the weather. Repeated entreaties to the developer were in vain. In most cases, he wouldn't even accept or return the buyer's telephone calls. After one year had passed, the residents of the community grew tired of being ignored by the developer and decided to exercise their rights as citizens by picketing the developer's latest business endeavor across the Potomac in Maryland. As soon as the developer learned of the picketing plans, he sought out the president of the civic association of the community, and begged that the picketing be delayed for only one week. This time, the developer remained true to his word. He started pouring the concrete on the following Monday morning. That incident occurred in 1966 and I still believe that but for the threat of picketing and the resultant deleterious effect it might have had on his development business, the developer may never have installed the streets. He certainly was not being pressured by Fairfax County, the government agency that should have pushed him from the start.

It is important for the potential new home buyer to understand that in most states, the builder can be held liable for construction errors for one year after a house is completed. If the construction error constitutes a structural deficiency, he can be held liable indefinitely.

Regardless of the effort that one puts into an investigation prior to signing a purchase offer for a new home, he can never be certain that he is receiving full value for his money. There are at least a hundred different ways that an unethical builder can camouflage sub-standard work when he is building a spec house. Unfortunately the majority of house buyers walk through a house under consideration, noticing only the floor plan, the paint, the walls, the appliances, and the visible plumbing. They have no way of detecting the flaws inside the walls, on the roof, or in the foundation.

FIVE: BASIC HOUSE CONSTRUCTION TERMS

Whether one purchases a new ready-built home, does his own subcontracting, or contracts with a professional builder, there are a number of basic construction terms that will serve him well at some point during the process. It will be helpful to be familiar with basic construction terminology when dealing with real estate developers, builders, or subcontractors, and even with building supply activities. Listed below are the basic terms, along with brief comments on the more important ones.

1. The foundation: Foundations are the parts of the house which support the entire weight of the rest of the house. Foundations must be located on well compacted soil or in concrete and must be strong enough to support the entire superstructure. They must also be resistant to the weather and to termites and other pests. Foundations have three main parts: the footings, the foundation wall, and the piers. Footings generally consist of concrete poured into small trapezoidal ditches dug in the ground and are usually six inches deep by sixteen inches wide. They must be installed below the frost line, because alternate freezing and thawing can cause excessive movement. The foundation wall rests on the footings and the tops of foundation walls become the surface on which the floor and superstructure rests. Foundation walls are normally made of concrete, concrete block, or some other masonry product. Walls of basements are normally eight inches thick. Houses without basements generally have foundation walls that are twelve inches thick. A pier is a column of masonry material, is generally of concrete block or brick, or a combination thereof, and supports the building in the space between foundation walls.

There may be a need for numerous piers. They also must rest upon concrete or some other durable, hard, and pest resistant material.

2. The basement and crawl space: Basements are prevalent in the northern part of the United States where the frost line is particularly deep and where terrain permits the incorporation of a basement into the house. They are relatively sparse in some of the southern states. However, because of today's mobile society, there are many who want basements in their houses, even in the southern states, because they have found that basements can serve many purposes other than simply providing foundation walls for a house. One needs to know, however, that not every lot will accommodate a basement. The soil may be too wet or too porous, or the lot may be too level to permit adequate drainage of a basement. For a basement, the foundation wall is built to a height of eight feet or more and a concrete slab is poured for the basement floor. The concrete will need to be thicker in those areas that are to support the weight of a brick fireplace or steel posts supporting the superstructure. Basements are generally more economical to build than comparable parts of the superstructure. In areas of the United States where frozen ground is not a problem, crawl spaces are quite common. Crawl space foundations are built similarly to basement foundations, except that the walls are not nearly as high as the walls of a basement. Foundation walls for a crawl space should be at least eighteen inches from the bottom of the floor joists to the ground. Also, in a crawl space, it is not necessary to pour a concrete slab. It has a dirt floor.

3. Cross Ventilation: Ventilation underneath the floor structure is necessary to prevent excess accumulation of moisture. To provide ventilation, it is necessary to construct vents in the foundation walls to provide for the cross flow of air underneath the house. In addition to the vents, heavy plastic sheeting should be laid over the ground in crawl space areas. The plastic prevents moisture from rising from the ground and permeating the flooring above.

4. Waterproofing: Basements and crawl spaces must be waterproofed to prevent water from seeping through the walls into the basement or into the crawl space area. The walls are waterproofed by applying layers of felt and asphalt to the walls. This system is sometimes called parging. A four inch perforated plastic or tile pipe is laid in a trench of several inches of gravel outside the footing. This permits the surface water to drain away from the foundation.

5. Framing: The frame of the house is like the skeleton of the human body. It is the basic component to which all other appendages are attached such as wallboard, siding, windows and doors. Lumber for the frame ranges from two inches to four inches in thickness and from four to twelve inches in width. The framing lumber is generally southern pine, fir, or spruce. For purposes of this book, the framing of a house will be discussed under the topical headings of floor framing, wall framing, and roof framing.

6. Floor Framing: The floor framing is the first framing on a house. The floor framing consists of sills, joists, girders, bridging and sub-flooring. The sill is the first member to be installed. It is two pieces of 2"x10" (sometimes 2"x12") lumber sandwiched together and placed on edge laterally on top of the foundation wall to provide a rest for the ends of the floor joists which are to be placed on it and perpendicular to it. The floor joists are 2"x10" or 2"x12" lumber placed parallel to each other with its centers 16" or 24" apart. They support the main load of the house. They usually do not span the complete width of the house, but the far end of the joists rests on top of girders which rest on top of piers under the house (or on steel posts in the basement). Girders are two or three pieces of lumber sandwiched together with their ends coming to rest on piers. Bridging is blocks of lumber that are installed between joists for latitudinal stability. After the sill, joists, and girders have been installed, the sub-floor which is generally 5/8" plywood is nailed and glued to the top of the joists in perpendicular direction. Later in the construction process, a layer of particle board may be installed on top of the plywood prior to eventual installation of hardwood flooring, tile or carpeting.

7. Wall Framing: This is probably the singularly most important phase of the framing process. All the rest of the house depends upon it being done accurately. If the upright two by fours are bent or bowed, the strength of the superstructure is diminished. Wallboard is difficult to attach properly, walls turn out not to be square, door and window alignment becomes more difficult, and problems may be created for the installation of plumbing, electricity, and heating/air conditioning. The main components of wall framing are plates, studs, and headers. Studs are vertical pieces of 2"x4" lumber placed 16 inches apart to which sheets of interior and exterior wall covering are applied. Some walls are load bearing. Some are not. The load bearing walls transmit the weight of the house above them to the sub-floor, which transmits the weight to the foundation. Studs are attached at the bottom to a sole plate (a 2"x"4" piece) that secures the wall to the floor. A double top plate is attached to the tops of the studs . This provides a reinforced area for connection of the ceiling joists and rafters to the wall . To cut out space for a window or door, one installs a header which consists of two pieces of lumber laminated together. Each of the timbers is generally six to ten inches in depth. The header spans the opening and provides both lateral and vertical strength for the wall, as well as helping to distribute the weight of the superstructure of the house.

8. Roof Framing: The roof design often dictates the method of framing that is to be employed. There are several different roof designs. The most common is the gable roof, then the hip roof and the shed roof. The gable roof is an inverted V. The hip roof has a mid-section shaped like an inverted V, but each end is also an inverted V slanted with the same pitch as the mid section to make each end section merge with the mid section. Another way to describe it is to say that the two mid sections are trapezoidal with the smaller of the two parallel sides at the top. The ends are triangular and sloped in such a fashion as to merge with the non-parallel sides of the trapezoid. The shed roof has only one side and slopes straight down from its highest to lowest point. Other styles are the Gambrel, the Mansard, and the flat roofs which are not described here. The slope of the roof is expressed as the pitch, which is defined as the number of inches of rise per foot of horizontal distance. An 8-12 pitch means that the roof rises 8 inches for every 12 inches of horizontal distance. The larger the rise per foot of horizontal distance; the steeper the slope. Major components of a roof are ceiling joists, rafters, ridge boards, hips, and valleys. The ceiling joists are normally 2"x8" pieces of lumber, parallel to one another. They support the load of the ceiling below and the roof above. Rafters are the diagonal members running from near the top of the walls to the top, or ridgeline of the house. Usually, they are 2"x6" pieces that support the sheathing (plywood covering) and the roofing. They are installed parallel to each other and connect to the ceiling joists at the bottom and to a ridge board at the top. The ridge board is at the peak of the roof and runs parallel to the long axis of the roof. Where two or more roof lines meet, hips and valleys are formed. A hip is the external angle formed by merging roof lines, and a valley is the internal angle that is formed.

9. Exterior Walls: The exterior walls consist of the sheathing and the siding. The sheathing is the first layer of outside covering that is nailed or glued to the wall studs of the house. It is a structural member because it strengthens the overall structure of the house by functioning as a horizontal and vertical brace for the walls. It also functions as an insulator because it resists air flow. With the rapid increase in technology there have been a number of developments in sheathing which tend to increase the structural and insulation capacity of the sheathing and , yet are available at economical prices. Foam sheathing has become more popular because of its insulation value, but it is not necessarily the strongest structurally.

Gypsum sheathing is relatively strong with some insulation value, but plywood is still recommended in areas that are subject to high winds. It is more expensive than the other types of insulation. Siding is the outside layer that covers the sheathing and provides additional protection from the elements. Sheathing can be both decorative and functional. There is a wide variety of siding in use today, including many types of wood, masonite, vinyl, aluminum, brick veneer, masonry other than brick, and stone. Wood siding is still one of the most popular. Today it is available in board and batten, square edged, and plywood. It can be obtained beveled, shiplapped and dropped. It is available in softwood or popular types of hardwood, with cedar being used a great deal for some of the more expensive homes.

Brick siding continues to be popular, because it is beautiful, long lasting, and comparatively maintenance free. Brick siding requires a ten inch thick foundation wall. The floor system rests on the inside four inches and the brick veneer rests on the outside four inches. An air space between the brick and stud wall provides an insulation barrier. Many people think of brick as an important load bearing part of the house. Actually, it is nonstructural. The loads of the house are supported by the stud walls, not by the brick veneer. The wide assortment of possible sidings cannot be described here. The type to be used is the prerogative of the individual after considering cost, esthetics, insulation and structural characteristics, and durability.

10. Windows and Exterior Doors: Both windows and doors are available in many different sizes and styles. All windows have basic components that are the same. The part that holds the glass is called a sash. Glass panes are divided by bars called muttons. The sash is surrounded by the window frame. The bottom of the frame is a sill, the sides are jambs, and the top is the head. In the case of the sliding window, the sash slides within the frame either vertically or horizontally. Vertical or horizontal windows are known respectively as double-hung and horizontal sliding windows. On swinging windows, sashes are hinged to the frame. Casement, jalousie, and hopper windows fall into this category. Fixed windows cannot be opened. They consist of picture, bow, and bay windows.

Exterior doors are generally 6'8" high and from 2'8" to 3' wide. Doors are made to provide as much weather resistance as possible.

They are generally of wood, either pine, birch, or oak, and they may be painted or stained. Some windows are now made of metal filled on the inside with insulation material. There are many different designs, but the most common doors are flush, panel, sliding glass, and french. The flush door has thin layers of plywood over a core of particle board. The panel door is characterized by a number of raised panels. Sliding glass and french doors are generally used as entrances to patios or decks.

11. Roofing: Roofing consists of roof sheathing (decking) and the roofing material. The sheathing is usually 1/2 inch plywood nailed perpendicularly to the rafters of the house. The most common roofing material is composition shingles, made of asphalt or fiberglas. Shingles are graded by their weight in terms of pounds per 100 square feet. The 240 pound grade is the most common. Some of the more expensive homes have wood shingles and shakes which are made of cedar, redwood and sometimes cyprus. They provide a more rustic appearance for a home, but it is controversial as to whether they are as durable or weather resistant as composition shingles. They are more expensive than even the best brand of composition shingles and their use as a roof commands higher insurance premiums because they are not as resistant to fire hazards.

12. Eaves: The eave is the part of the roof that extends beyond the walls of the house. The ends of the rafters are covered with an eight to ten inch wide piece of lumber that is called a fascia. A soffit board is used to cover the open space between the fascia and the wall. Plywood or one inch lumber is generally used for the soffit.

13. Insulation: Several different types of insulation are available. Batts or blankets sold in rolls are commonly used for wall,ceiling, and floor insulation. They can be purchased in widths of 16 or 24 inches to fit stud spacings of those dimensions and they are available in different thicknesses ranging from one to twelve inches. Most batts and blankets are made of fiber glass or rock wool. Insulation can also be obtained in loose form to be poured or blown into attic spaces. Insulation is measured by its resistance to heat flow. This resistance is expressed as an R-value. The higher the R-value; the more resistant it is to heat flow. In many states, floors are required to have a value of R-11 and ceilings R-19 to R30 for best results. To complement and facilitate the insulation,it is important that areas around windows and doors be considered for weather-stripping and that caulking be applied to openings around plumbing and wiring. Most homes tend to generate water vapor from bathing,cooking, laundering and other sources. This vapor tends to condense within the wall cavity, thereby creating a need for a vapor barrier to be installed on the inside covering of the insulation to prevent the passage of vapor. This barrier may be of plastic,foil, or kraft paper. Also, efficient attic ventilation reduces the possibility of condensation and decay of structural roof members. It also helps to control temperatures in the attic in the summertime, thereby reducing cooling costs for the rest of the house. Proper ventilation can be achieved by using gable vents, ridge vents, soffit vents or power ventilators or attic fans.

14. Walls and Ceilings: The interior walls are commonly finished by installing either 4 foot by 8 foot or 4 foot by 12 foot sheetrock on the interior stud walls. The 4'x12' sheets are installed horizontally; the 4'x8' sheets are installed vertically. If you do any of the work yourself, you will use the smaller 4"x 8' sheets which are much easier to manage. The joints are covered with wallboard tape that is applied with wallboard joint compound and smoothed on with a wallboard trowel. After applying the tape, the seam is permitted to dry for two or three days before applying the first coat of joint compound which is designed to cover the tape and fill the indentation where the two pieces of wallboard have been joined together. Again, it is necessary to let it dry for two or three days before applying the final coat of joint compound. If the final coat is applied by one experienced in the trade, it will not need to be sanded. If not, sanding of the joint may be necessary to make it meld evenly and unobtrusively with the flow of the wallboard. After sanding, the wallboard may be painted or papered. Most ceilings are of sheetrock that is either painted or sprayed with a textured paint designed to cover any imperfections that may remain after the installation process.

15. Flooring: There are a number of different type floors that can be installed over the sub-flooring, depending on the area of the house where the floor is located. In living room areas, you have a wide choice of different tongue and groove or parquet hardwoods, or you may choose to install carpeting over a plywood base. In the kitchen, there is a wide array of vinyls and of tile or hardwood. For entrance foyers, there is hardwood flooring, either tongue and groove or parquet, and there is tile or marble. Vinyl sheets are commonly used in areas of the house where there is apt to be water spillage or temporary deposits of residue which would tend to mar carpeting. Tile can be used in the same areas, but you may want to use tile very selectively because it is much more expensive than vinyl.

16. Molding: Molding is used to cover openings between the wall and ceiling, between the wall and floor, around all doors and windows and in other areas where something has been installed in or on the wall, such as a fireplace mantle, a wet bar, or book shelves. It is usually made of pine which comes in unblemished types or fingerjoint where it has been joined together with glue throughout its length. The unblemished type may be stained, but the fingerjoint type must be painted to cover the area whether it is fingerjointed together. In addition to the pine variety, you can purchase a more expensive type in oak or birch, which lend themselves readily to staining. However, there is no point in selecting the more expensive molding if you intend to paint it.

      17. Heating Systems: There are a number of different types of heating systems including the central furnace which uses oil or gas, and sometimes wood; the hot water and steam systems which distribute heat by means of radiators; the forced warm air system which employs gas, oil or heat pump to generate the heat which is then forced into ducts for distribution; electric baseboard heat; the space heater; solar heating systems; and thermal systems which extract the heat from underground sources. The most commonly used system today is the forced warm air system employing oil, gas, or electric furnaces in the colder regions of the United States,and a predominance of heat pumps in many of the southern states. The heat pump can extract heat from the outside air as long as the outside temperature is above 32 degrees Fahrenheit. When the outside temperature drops below freezing, the system employs a back-up source of heat that is provided by heat strips which are heated directly by electricity.

      18. Air Conditioning Systems: The most common air conditioning systems are central units generally mounted on a concrete pad outside the house, powered by electricity,using a system of ducts to distribute cool air throughout the house in the same manner that heat is distributed by the central forced air system. The same duct work generally serves both the central heating system and the air conditioning system.

SIX: SUBCONTRACTING YOUR OWN HOME

We have already discussed three options that one might consider when electing to acquire a new home. Those options are the fixed price contract, the cost-plus-fixed-fee contract,and purchase of a new ready built home. There is a fourth option that is being selected by an increasing number of home buyers each year. That is the option of sub contracting the building of their own home. When one elects this option, he decides to do all the managerial work himself, obtain his own financing, hire and supervise the subcontractors, and order and assure timely delivery of all the material not provided by the subcontractors.

Traditionally, people have been reluctant to assume responsibility for the functions associated with subcontracting the building of their home, yet those who have been willing to accept the challenge report that the job was not nearly as difficult as they had anticipated and nearly all of them report a sizeable savings as a result of having performed their own subcontracting. It is important for one to realize at the outset of planning that to do his own subcontracting, he doesn't have to be a professional builder, doesn't have to be familiar with the building code, and doesn't have to be a carpenter, plumber, framer, electrician, painter, roofer, or any of the allied tradesman who work on a house. The endeavor requires nothing more than common sense and the willingness to spend a considerable amount of time in selecting the right subcontractors. Not even the most experienced builders are expert or even knowledgeable in all the tasks associated with the building of a house. They depend upon each subcontractor in much the same manner that the ordinary citizen depends upon any type of service personnel in the process of daily business. Also,there is an additional comforting factor that the work of each subcontractor must be inspected and approved by the county or city building inspector. If the work is unsatisfactory, or not in conformity with the code, the building inspector will require the subcontractor to remedy or redo the faulty work. There are however, a number of areas in which you, the owner, must develop a degree of knowledge and be willing to exercise a high degree of diligence. The knowledge and diligence required are well within the scope of a large percentage of the citizenry who at the inception of such an effort have virtually no knowledge of the building trade.

After you have obtained the plans and specifications provided by the architect, have purchased the lot of your choice, and are ready to start building, you need to obtain a building permit for the construction of your home. To do this, you need to visit the county or city building inspectors' office, where you can obtain information on the required building standards and a copy of the local building code. The inspectors office may ask for certain documentation in order to advise you properly. For instance, they will probably ask for a copy of your plans and for a survey showing the proposed location of the building on the lot that you have selected. To avoid extra travel, it is wise to call the building inspectors office before visiting, so that you may take any required documentation with you on your first trip. The inspector will inform you whether your plans are in conformity with the zoning of the area in which you intend to build and whether or not you may need a variance that may have to be approved by a board of appeals. After final examination of your plans, the building inspector will either issue you a permit or notify you of any changes that may be required to comply with the zoning or the local building codes. These changes may have to do with where your house is to be located on the lot, the minimum square footage required for the building, or the types of materials that you propose to incorporate into the building. After you comply with his requests the permit will be issued. You will have to pay a fee for the permit. This fee is normally in the vicinity of three to five percent of the total construction costs. In some localities the fee that you pay at the outset includes the costs of all inspection fees throughout the construction process, such as the inspections for the foundation, the framing, the boxing, the plumbing, the heating, and the electrical work. In other localities, you will pay a smaller fee at the inception, but will pay a fee each time in the future that the inspector visits your house. After obtaining the permit, you post it on a sign in front of the future house and you are almost ready to hire your first subcontractor. Before hiring him however, you will need to set up the sequence of tasks that will require different types of subcontractors, and you will need to formulate your procedures for obtaining a reliable subcontractor for each task at the best price that you can obtain.

In the general order of performance, the tasks that are normally associated with the construction of a house are the excavation, the footings, the foundation, the framing, the fireplaces, the roofing, the plumbing, the electrical wiring, the heating and air conditioning duct work, the sub-flooring, the flooring, the boxing, the exterior masonry, the insulation, the dry wall, the trimming, the well digging, the septic system, interior and exterior painting, grading and leveling, the shrubbery, raking and sowing, and cleaning. Some of the above may be done simultaneously, but it is important to plan in such a manner that each task be completed in a timely manner so that tasks which depend upon completion of a previous one can begin as scheduled. For instance, the dry wall cannot be installed until the insulation task is completed, and the insulation cannot be installed until the electrical, plumbing and heating tasks have been completed. Nor can the dry wall, cabinetry, and flooring be installed until the house is "dried-in"; i.e., the roofing installed and windows and doors set.

If you are thinking of managing the subcontracting of your house, it is essential that you become familiar with the plans. This does not mean that you have to be knowledgeable in how the plans are executed. It means simply that you must be familiar with what is to be done, where it is to be done, and the type of material that will be used. How it is to be done is the job of the subcontractor. The most formidable task that you will be required to accomplish will be that of hiring the subcontractors.

You will be ahead of the game if you are fortunate enough to have prior knowledge of the prevailing rate of pay for specific skills such as carpentry, plumbing, deep well drilling and/or other skills. However, even if you have no prior experience in any of the subcontracting tasks, you can still be successful in your subcontracting endeavor if you are willing to spend the necessary time and effort to learn possible sources of good subcontractors and the prevailing rates for their work.

As the time approaches for a specific task to be performed, you should start your search for the names of qualified subcontractors. At the same time, you should be educating yourself on the prevailing rates for subcontract work of the type you are considering. Many builders will be glad to tell you the rates that they pay for the particular type work. Many of them will also be glad to tell you the names of subcontractors that they know who work at reasonable rates. Building supply activities will also be able to recommend subcontractors who have traditionally bought their supplies from them. Real estate agents frequently have knowledge of subcontractors who have performed work at reasonable rates on projects that have been sponsored by their agency. The county or city building inspector will have an unlimited source of subcontractors as a result of having inspected the work of each at other houses that have recently been constructed. In my own case, I announced at a meeting of the local Rotary Club that I was badly in need of skilled masons. At the end of the meeting, my fellow members provided me with five recommendations.

Prior to hiring subcontractors, you will want to obtain competitive bids from at least the top three that you have selected as potential candidates. If you have carefully performed your investigation, you will have a good idea of what is to be considered a fair quotation, and you will be able to judge the bids accordingly. Once you and the subcontractor have agreed on the fee for his services, you will need only to show him the plans. If you don't consider yourself sufficiently qualified to evaluate the quality of his work, you may elect to entrust the evaluation to the building inspector who will not approve the work until it is satisfactory. However, you are cautioned that the Building Inspector is required only to assure that the work is in conformity with the code. He is not necessarily required to assure that the work is in conformity with specifications. In those cases where the specifications call for work exceeding code requirements and you feel that you are not capable to evaluate the work, you may want to call in an independent evaluator, such as a skilled carpenter, plumber, electrician or other skill, depending upon the type of work to be evaluated. Another alternative is to call another builder to act as a consultant for a one time fee.

If you elect to do your own subcontracting you may want to hire a general foreman to assist you in obtaining subcontractors, in determining prevailing rates, and in evaluating the work of subcontractors. The foreman may not have to spend more than an hour or so each day at the job site and he may be willing to work at an hourly rate that is not appreciably higher than the hourly rate of some of the skilled laborers employed by your subcontractors. Such a foreman will more than likely be an individual who has spent several years working on building projects, but has not yet obtained a builders license. Again, the various building supply activities may be a valuable source of information for the possible names of such foremen. Another option is that of employing consultants to assist you only when you encounter specific problems that you feel that you do not have the ability to address. Strangely, this is one of the most economical routes that you can take. When you contract with a builder, you will pay him directly or indirectly, a fee that is somewhere between 25% and 30% of total building costs. For a $250,000 dollar house, that fee equates to a range of $62,500 to $75,000. If during your own course of handling subcontracting, you have to call in a consultant as many as ten times, you will pay them somewhere in the range of $2,000 to $3,000. It has been my experience that consultants are relatively easy to find and they are generally eager to earn an extra income of $150 to $250 for only two or three hours of their time. Virtually all builders are potential consultants, that is unless they have been so prosperous that they are not interested in earning a few hundred dollars on the side.

When my own house was undergoing construction, I employed a

structural engineer who spent two hours reviewing the construction in my house and two hours formulating remedial plans and directions. His total fee was $190.00, even though his effort provided the entire basis and direction for completing the rest of the work on my house. When the house was almost finished, a question arose as to the bearing load that an existing support beam could carry. On that occasion, the consultant evaluated the situation and provided remedial directions for a fee of $45.00. There were only two other occasions when I had the need for a consultant. On each of those occasions, the consultant fee was $275.00, even though each occasion required the formulation of a written report that consumed approximately four hours in preparation. One may ask why a builder would be willing to work as a consultant for such low fees. The answer is that, as a consultant, he has nothing at risk. He is gratified that he has been sought out for technical advice. The fee is welcome as an additional source of income and the good will created may pay off in future business. For the individual building his own house, the consultant can spend only three or four hours and provide enough direction to go a long ways toward ultimate completion of the house. However, if one is contemplating the subcontractor option, there are two additional items of which he should be aware: the need for him to remain alert to the possibility of theft from the building site, and the need to know that there is a tendency for many subcontractors to inflate their bills. The public is generally unaware of the tremendous losses to theft which occur daily within the residential construction industry. The largest percentage of such thieves, know what material is stored at the site, and visit it at night to help themselves.

There is not a lot that can be done to prevent construction industry theft, except for the builder to assure that he never has too many valuable or attractive items on site until time comes for their incorporation into the building.

When a construction site is inactive for a period of several weeks or months, thieves descend upon it like vultures descending upon a dead carcass.

Many subcontractors have become adept in the art of "add-ons", a procedure whereby they add an amount in the "total due" column of their bill for extra work which they claim not to have expected when they accepted the job. In one case, a subflooring contractor added fifty dollars because he swept the room before starting his work. In another case, a floor tile contractor claimed that he had installed seventy five square feet of tile more than he had actually installed. In another, a roofer wanted seventy five dollars for installing a three foot rain diverter which had been listed in the specifications upon which he had bid. There were numerous other cases. You need to be alert to this ploy, because in most cases, subcontractors regard it as a normal thing to attempt to maximize their income for work performed. Most of them will smile and delete the add-on when questioned.

In Chapter 2 we began the narrative portion of the chronological sequence of adversities that accrued to my wife and me when we rushed into a building contract without first doing the homework that would have prevented those adversities. Chapter 7, "The Beginning of the Nightmare", continues the personal experience saga and is devoted to the first several months that we were under a contract with Beryl Jaker. For those whose itinerary may preclude devoting the time to the reading of this section, the reader may continue directly with the substantive matter of the book by starting anew with Chapter 8.

SEVEN: THE BEGINNING OF THE NIGHTMARE

January 1986 should have been one of the happiest months of our lives. I had completed my professional career and could now look forward to the many benefits of retirement. We had just signed a contract for the construction of our retirement home and were looking forward to seeing it take shape and to assume the characteristics of all the plans that we had made for so many years. We didn't have the slightest idea at that moment that our plans were turning into a terrible nightmare from which we would awaken only after two years of taxing every ounce of strength and ingenuity that we could muster. We didn't realize that our earlier dreams were in direct conflict with the goals of the man whom we had trusted to build our home. Our ordeals started from almost the very moment that the contract was signed.

Three weeks after awarding Mr. Jaker the contract, he finally informed me that he would start excavating the basement the following week. After he started work on the site, excavation was completed in about three days but at a cost that was approximately 25% more than that paid by other builders for similar work. Mr. Jaker had failed on his first job to notify me of the projected cost as promised, or to obtain competitive bids. It was only a few weeks later that one subcontractor informed me that Mr. Jaker was volunteering to pay fees exceeding the usual in return for his receiving partial rebate of the overpayment.

Following completion of the excavation, another three weeks elapsed before Mr. Jaker installed the footings. Then, they were disapproved by the building inspector because they were not in conformity with the code. When I queried Mr. Jaker as to why the work had not been done properly, he replied that subcontractors knew their work, so it wasn't necessary for him to be there. After the footings were poured, the site remained idle for another three weeks before a mason arrived to start installing the foundation. This mason soon established a pattern of working one day and not returning to the site until a week later when he would work one day and again absent himself for a week. I finally learned from other sources that the mason was working full time on another job and was reporting to Mr. Jaker's job only when there was a one day break in his regular job. Jaker informed me the next day that he had fired the errant mason in accordance with my wishes, but stated that he didn't know where he could find another.

A period of six weeks went by without any work being done at the site. Each time I queried Mr. Jaker, he informed me that he was still trying to find a mason. Finally, in desperation, I announced at one of my Rotary Club meetings that I was in dire need of masons. Several members responded with names and telephone numbers of masons who were looking for work. Jaker hired a crew from the list but informed me that the cost was going to be $300.00 per thousand bricks. This was at a time when other builders in the area were paying $160.00 per one thousand bricks. I informed Mr. Jaker that I had been naive, but that I wasn't that naive. He made a special trip to my house that night to inform me that he had talked the crew into laying the brick for a mere $200.00 per thousand.

My wife and I had selected bessemer gray as the color of bricks for our home, but Jaker ordered a biege color which would in no way conform with the color scheme that we had selected. When my wife and I discovered his error, the job had proceeded too far to tear down all the brick, so we reconciled ourselves to the fact that we could not have the color of our choice and started making plans to change the overall color scheme for the entire building. Then Jaker compounded his mistake.

On his next order, he requested the bessemer gray brick that he should have ordered the first time. The crew was installing the brick around the base of the garage when a crew member pointed out to me that they had laid about 500 of the mismatched brick. When confronted with his latest error, Mr. Jaker's response was that the mismatched brick was close enough to ground level that he could have the dozer operator heap dirt up next to the house and gradually contour the fill away from the house in such a manner as to bury the gray brick. It was apparent that he was not at all concerned about the poor quality of work going into the house nor of accompanying costs. My wife suggested that if we didn't rid ourselves of him, we could easily pour our life's savings into the house and wind up with a home not remotely resembling the house that we had planned. I replied that, at all costs, I wanted to avoid getting the house tied up in the courts. We didn't know the half of it. He was to make numerous additional costly mistakes, mainly because of his failure to instruct the workers properly, or to follow the plans.

On the day that the concrete was poured for the 1800 square foot basement, Jaker let the crew start pouring the concrete at 1:00 o'clock P.M. knowing that the weather was threatening rain, and that the job would require ten to twelve hours to complete. He had an engagement to take his family to the beach over the week-end, so promptly at 6.00 o'clock, he left the job and had the audacity to ask my wife and me to look in on the crew doing the work. We stayed with the crew until midnight, during which time we made several trips to obtain electric cord and lights so the crew could see what they were doing. It was during this time that the lady in charge of the crew informed us that Jaker had propositioned her to overbill for the job and give him a kick-back.

It was becoming apparent that Mr. Jaker was interested only in the funds that he could draw against the house and in the kick-backs that he could generate among suppliers and subcontractors. This became even more apparent when I reviewed some of the invoices which accompanied lumber deliveries to the site. At that time, construction grade 2 x 4's were being sold by nearly all lumber dealers for $1.48 per 2 x 4. Mr. Jaker was paying $1.79. From time to time, I noticed the supplier was delivering sub-standard material or was breaking an excessive amount of material by its being dumped from the delivery truck to the ground. When I queried Jaker about these losses, he informed me that in his many years of building, he had never returned anything to the vendor. He said that I worried needlessly: that in the end, everything would total out at about the price originally estimated.

When the basement was finally poured and the brick work around the bottom part of the building had been laid, a period of six months had elapsed since Jaker had been awarded the contract. The foundation still was not completed and the floor joists supporting the entire super structure of the house were still resting on 2 x 4's laminated together to support the weight of the house. As the foundation was not completed, Mr. Jaker still had seven more months to complete the house and still be in conformity with the contract. He was not at all reluctant to display his anger when either of us inquired why work was not progressing faster. Nevertheless, with our constant prodding, he hired a framing crew and started framing the super structure of the building. About the fourth day of their work, the foreman complained to me that he did not always know what he was supposed to do; that Mr. Jaker had given him only broad general instructions and had provided him with no plans. He said that Mr. Jaker was visiting the job site only about every three or four days. When the framing was about sixty percent completed, my wife became very much alarmed when she observed that all the roofing rafters had been improperly installed, and if left in their present configuration would so change the profile of the front of the house that everything would be lost.

Figure 1: The above photo shows the rafters as installed before dismantling.

Figure 2: The above photo shows building after faulty rafters were dismantled

Jaker agreed that the error had been made and that the roofing rafters would have to be dismantled, the pitch of the roof would have to be changed, and new rafters installed. He said that he thought he could still use the old rafters.

About midway through the framing phase of construction, I noticed that a significant number of studs were not properly plumbed. That is to say that the upright 2 x 4's which collectively form the framework for the wallboard were listing from the vertical position, in some cases as much as two inches off the vertical in eight feet of height. I marked each with a blue crayon with a short message to Mr. Jaker, who ignored the messages. When I asked why he had taken no corrective action, he replied that he had not seen the messages and, that in any case, the errors would be covered when the wall board was installed.

About the same general time, I discovered that the framing crew had cut the stairwell opening and fireplace opening ten and one-half inches too far toward the front of the house. This was an error with major implications, for it changed the layout of the entire first floor. Jaker casually remarked, "Well, as I see it, you have two choices. We tear out the floor and we do it again, or we leave it as it is and make all the rooms on the front side of the house ten and one-half inches more narrow". After considering the expense, Becky and I reluctantly decided to reduce the size of the forward rooms.

Every aspect of the construction that had been completed had exceeded Jaker's quoted estimates by a factor of at least twenty-five percent. While I was still discussing the error with Jaker, Becky observed that the window opening in the laundry room had been installed in such a manner that the window, if not corrected, would provide a view of the inside part of the roof overhang instead of a view of the front lawn, as provided by the plans. When shown this error, Jaker instructed the framing crew to change it, paying an extra $25.00 for the job. Of course, the expense was charged to me. We were beginning to wonder if we would ever have a home that remotely resembled the home that we had so enthusiastically envisioned at the beginning of construction. In fact, everything that Jaker said or did convinced us that we were on a downhill roller coaster with no brakes.

After the framing crew completed their work, we had a short conversation with the foreman who revealed that Mr. Jaker had met the foreman in a parking lot and upon Jaker's discovery that his new acquaintance was a framer, had hired the foreman immediately to frame our house. Jaker had no knowledge of the framer's background or his capability to meet minimum standards. It developed that the framer and his crew had framed a few houses in the $60,000 to $80,000 range, but had never had any experience on houses in the $200,000 to $300,000 range. The results of the framing showed the inexperience of the crew. Within three days after their departure, all of the roofing felt had either blown off the roof or it was loose and flapping in the breeze like so much confetti during a Broadway parade.

Figure 3: Front view of the roof with the roofing felt blown loose

Figure 4: Rear view of the building with the roofing felt blown loose

With the roof bare, large one inch cracks were revealed between all the plywood sheathing on the roof. Jaker argued that roofing felt on a house was unnecessary, and that the best approach was to install the roofing with no felt underneath. I informed him that all houses had roofing felt; that its use was common to all builders, and that I wanted the roofing felt replaced. He took no action at all, but continued to procrastinate.

I went to the building site early on a Monday morning and found a roofer attempting to install shingles over the bare plywood. He asked if I were the man for whom the house was being built. I informed him that I was. He said, "Sir, it is impossible to put shingles on this building in its present condition. There are no rafters to support the edges of the roofing all the way around the house, the fly rafters have not been installed, and the cracks between the pieces of plywood sheathing need to be filled. I'll be honest with you, I'd rather not do the job than attempt it with the felt and sheathing in such a state." I told him that I appreciated his honesty, and that by all means he should discontinue the job.

During this same period, Jaker had engaged a bricklayer to construct the fireplace and chimney and he had hired a sheetmetal crew to install the ductwork for the heating and air conditioning system. All arrived and started work during Jaker's absence from the vicinity. Troubles began to compound themselves. The bricklayer pointed to the cathedral ceiling in the family room and informed me that the roof at that point was not strong enough to support bricklayers or the bricks for the chimney which would have to be lifted up onto the roof. He had to erect a temporary support in the middle of the great room to reinforce the roof at the weak point, thereby enabling him to finish the job. I then examined the roof in the living room. It had the same defect. Jaker had modified the design of the building without the educational background to compute the stress at the point where the opposite members joined, and had created a condition that was flagrantly unsafe, and one that he was powerless to correct.

The sheetmetal crew had arrived without instructions and was proceeding by guesswork to install the various vents and returns throughout the house without any regard for functionality or esthetics. We asked them to stop work until we could get together with Jaker and formulate some type of disciplined approach. The plumbers' efforts were equally disasterous. The chief of the crew was in the hospital, so the crew came to do the work without their boss. They installed the effluent pipe for the entire system in such a manner as to deny access to the basement stairs to any person more than six feet tall. This was in violation of the local code which specifies that all such pipes must afford a minimum overhead clearance of six feet and ten inches. The rest of the system was a mish-mesh of pipe connections and misrouted pipes without adequate numbers of clean-out valves and without proper gravity flow. We also asked this crew to discontinue their work until we could get Jaker to help solve the many problems that he had created. On several occasions, he told me that I would have to trust him; that he knew construction and I didn't. I finally told him that I sincerely wished that were the case; that I had employed him because I thought he knew more about construction than I, but events were now proving that assumption incorrect. It appeared that even as a novice, I knew more about construction than he.

Things began to reach a boiling point. The roof was bare of felt. The basement contained at least a foot of water. The edges of the roof were drooping like pieces of wet cloth. There were sags in the roof where Jaker had failed to see the requirement for reinforcement underneath. The lot looked like a huge junk yard with piles of lumber that had been cut wrong and tossed to the ground. Instead of things improving, they were deteriorating. Becky and I agreed that Jaker's contract had to be terminated. If his conduct here did not constitute non-performance of his contract, I didn't know what would. I called and asked Mr. Jaker to come to my house. I explained to him that we had been very dissatisfied with his work; that in my opinion, he hadn't tried to do a good job; and that he had resisted everything that we had ever asked him to do. I then outlined the many things wrong with the house. He said, "Perhaps I have spent too much time on the March house. Give me ten more days and I will straighten up everything to your satisfaction." I doubted that Jaker could straighten out his mess in ten days, but for the sake of fairness, I told him that he could have ten days during which time he was on probation.

Granting him another ten days was a poor decision. His first action after our meeting was to go to the bank and borrow an additional $25,000 against our house. He had already borrowed funds far in excess of the materials and labor that had gone into the house. We were to learn later that he had drawn more than $100,000 against the house and could produce receipts for only $18,000 paid out for labor and supplies.

Jaker turned up at the job site on Monday morning with a crew of three carpenters. They worked for about four days, then all of them disappeared with no more than ten percent of the faults corrected. The following week, only the roofer reported to the job site. Jaker had told him to proceed with installing heavy timberline shingles on the roof over the great room, the area that was not strong enough to support itself, and would collapse with any added weight. The roofer told Jaker, "Mr. Nash told me that he did not want roofing installed in that area until some type of support is installed underneath." Jaker replied, "To Hell with what Mr. Nash says. I am the builder and I want it installed". Jaker left the scene and the roofer came down and told me of his conversation with Jaker. He added, "Personally, I am afraid to work in that area". I instructed him to stop his work and told him that I would call him when I was ready for him to start again. I then walked into the house and noticed immediately that the weight of the roof had pushed a 2 x 4 bracing from its position at the peak of the cathedral ceiling. We decided there and then to fire Jaker.

I called my attorney and informed him that I was about to terminate Jaker's contract. He said, "Call me after you talk with him, if you think you will need my help." I said, "First, I am going to call in a structural engineer and ask for a written report of deficiencies."

I accompanied the engineer to the site. His report stated that the building was in terrible structural condition, but not beyond salvage. His report identified a number of specific reinforcing actions needed and portrayed in drawings and sketches the details of the actions to be completed.

On Monday, I told Jaker that the report was not good and that we had decided to terminate his contract. He said, "You can't." I said, "We feel that we can, in view of the fact that your effort has been one of non-performance from the inception of the contract." He replied, "I'll finish the house and sell it." I said, "I'll get an injunction if you so much as set foot upon that property." He left as angry as a threatened hornet.

Three days later, friends notified us that they had recently had occasion to ascertain whether Jaker was a licensed builder. Their search revealed that he was not licensed. As an unlicensed operator, he was not legally authorized to enter into a building contract, the total value of which exceeded $30,000. So after Jaker departed my house with his angry quote, I immediately called our attorney and informed him that Jaker was unlicensed.

We had been thrust into a position whereby we could easily lose our life's savings, but even then, we were not aware of the magnitude of the financial damage that could be inflicted upon us, all because we had innocently and trustingly engaged a builder without checking out the fundamentals that have now been outlined in this book.

After I informed our attorney of our latest conversation with Jaker, he wrote Jaker a letter informing him that his contract was terminated; that contracting without a license to build houses in the state was a criminal offense; and that under no circumstances was he to obligate any more funds against the building. Jaker retained his own lawyer. Among other things, Jaker soon learned that it was not exactly a criminal offense to contract without a license, but having no license did mean that his contract was invalid. In my case, the chief difficulty was that Jaker and his wife had borrowed money against the property and our attorney, David Hoffer didn't have any ready answers as to how to get our property back into our names. David quickly learned that Jaker had not only spent the $30,000 which I had advanced him, he had drawn an additional $71,000 from the bank. The bank therefore, had a $71,000 lien against my property. Jaker had spent $101,000 from funds drawn against our house and could produce paid receipts for only $18,000 of the $101,000. Clearly, he had been using the money generated by our house to pay his other debts. Jaker claimed that he could produce only $10,000 in cash. He had obtained from the bank a total loan authorization of $164,000. At the time that he was fired, the building was less than thirty percent complete; hence the bank should not have advanced him more than thirty percent of $164,000, or $42,000. Yet they had advanced him $71,000. David said that it would be foolish for us to put any more money into the house while it was in Jaker's name; that the lending agency may have to foreclose to get its money back; and, if that happened, we would simply be contributing more money to the bank. He told us that he believed that the bank had failed to exercise due diligence in lending Jaker too much money, and he asked our permission to employ a legal data bank in Charlottesville, Virginia to make a data search to ascertain whether there had been similar cases against lending institutions elsewhere. He said that the data search would cost about $600. When the report came, it was inconclusive. There had never been a similar suit in North Carolina. There had been one in South Carolina, where the court had ruled that banks have a moral obligation to protect the public interest by being diligent in their actions when a third party interest is at stake. There had been cases in California where the courts ruled that the rights of the third party had been subrogated. A case in New York had resulted in a ruling against the bank. Our dilemma rested in the fact that pursuing the matter through the courts would be time-consuming and expensive. It was estimated that legal expenses would be somewhere between $2,000 and $5,000, and we could not be sure of the outcome. Legal action would mean that the property would be in limbo for a period of one to two years. There was no question but what the bank had been derelict in considering my right as the third party. The bank knew that Jaker had obtained the lot from a third party and that the third party had temporarily deeded the lot to Jaker as part of the building contract. In spite of this knowledge, the bank advanced Jaker more money than his progress in building justified. Almost any jury would have ruled for the third party bringing such a suit against a bank, but it is a gamble anytime that a citizen goes to court, regardless of the justification for his suit. It was a gamble that we could not afford.

I had agonized over David Hoffer's advice. He couldn't get Jaker to deed the house back to us, yet he wanted the property to continue to deteriorate until routine legal matters took their course. Eventually it dawned on me that David's course of action was impractical. I called Jaker the following morning to see if I could reason with him. I said, "Beryl, I have an appointment with the District Attorney tomorrow morning. I am going to talk to him about criminal prosecution. I thought you might want to meet with me and talk about the matter." I was surprised when he readily agreed to meet with me. It occurred to me that even though he thought he was immune to criminal prosecution, he wasn't certain; therefore he thought he had better do what he could to avoid having the D.A. look into his activities.

At the meeting, Jaker agreed to provide lien waivers signed by a number of his suppliers and subcontractors and to sign a promissory note for $31,000, whereby he agreed to pay me $3,000 per month from money that he was to draw from the bank to construct the other house that he was building. He also agreed to petition the bank to provide the records of loan activity portraying the inspection reports and the dates they had made monetary advances to Jaker against the progress of construction.

Our attorney prepared the promissory note for the signatures of Jaker and his wife. The note specified that he was to pay me $6,000 immediately, and $3000 per month thereafter. Jaker signed the note, but informed me that his wife would refuse to sign it, so I accepted the note with only Jaker's signature on it. He would immediately put every piece of property that he owned into the name of his wife.

David Hoffer still felt that we should sue the bank. I told him, "David, there was no other way. If I had followed your advice and then lost the case, we would have been destroyed. We couldn't afford to wait while the legal system took its course and our house continued to deteriorate."
The following day I started following all the leads that I could find in search of an attorney with whom we would feel comfortable in bringing suit against the bank. We made a surprising discovery, the revelation that practically all the good lawyers and most of the mediocre ones were in the pockets of the lending institutions. Lead after lead became worthless when each attorney that I called, boldly informed me that he couldn't take the case because it would entail conflict of interest; that the lending institution regularly patronized their offices with various types of administrative and legal assignments and that if he should represent anyone in an action against the bank, it would be the end of all business which the institution regularly provided him or his firm. Finally, one of the attorneys who had been recommended also said, "I am one of those attorneys who would have a conflict of interest, but I have an attorney acquaintance, whom I can highly recommend. I know for a fact that he has no such conflict of interest. He is really like a tiger. When you give him a problem, he digs in with the tenacity of a bulldog. His name is Haywood Coward. I suggest you give him a call."

The nightmare was reaching a high crescendo. It was necessary for my wife and me to spend $71,000 to get property returned to us that already belonged to us. There seemed to be no refuge from the storm. We hadn't been able to find an attorney to represent us against a major lending institution, even though our case was a just one. Everywhere we turned, there was a demand for more money. We had signed a contract with a real con artist and the court records verified many of his misdeeds, yet his credit worthiness was undiminished. In the face of everything that he had done, a major bank was issuing him a new loan in the amount of $183,000 to build a new house. I couldn't help but wonder what Shakespeare could have done with such a scenario. The nightmare couldn't get worse, it had to get better. The end of the nightmare is related in Chapter 12. We shall now return to the substantive instructions for contracting for the building of a home. The next chapter deals with obtaining mortgage money.

EIGHT: OBTAINING MORTGAGE MONEY

At some time during the construction process or immediately thereafter, most new home owners obtain mortgages on their property. The mortgage money may be needed for a variety of reasons, but the most common reason of all is the fact that mortgage interest provides the owner a very important and very necessary tax deduction. There are certain fundamentals of mortgage acquisition that are just as important to ones financial security as any other facet of the building process. One can benefit by having advance knowledge of how the mortgage process works.

Try to obtain the construction loan and the permanent loan at the same time. If one elects to subcontract the building of his home, he will need to obtain his own construction loan. If he elects to engage a general contractor, he may or may not choose to obtain his own construction loan, depending upon the route that appears to be the most practical. A construction loan is one whereby the lender advances funds at various stages of house completion. Normally, the lender will expect the construction loan to be paid as soon as the house is sold or when final closing takes place. Of course, when you go to final closing, you will need a permanent loan, one for 15, 25, or 30 years depending upon your preferences for amortizing it and the terms associated with each of the three. It is better if one plans far enough ahead to have the lending agency consider both loans at one time, or in other words, consider the construction loan first followed by the permanent loan. There are a number of factors to consider when applying for the permanent loan. The importance of each factor may vary according to the tastes and the financial situation of each applicant. The cost of the loan is determined over its lifetime by the amount of the principal, the interest rate, and the points paid to obtain the mortgage at the particular rate. A point is the equivalent of one percentage point of the loan. It is paid up front, or at the inception of the mortgage and is a one time payment. It does not have to be paid anymore after the initial payment, and like interest on a mortgage, it is tax deductible, but only on the initial mortgage. It is not tax deductible on a refinance. In addition to the interest rate, the term of years, and the points, the applicant must consider the settlement costs, which in some localities can be very costly. Settlement costs include attorney fees, appraisal fees, loan origination fees, inspection fees, deed recording fees, insurance, notary fees, surety bonds,and in some communities, flood control insurance fees. If you have two settlements, you may have to pay all these fees twice. It is therefore prudent to obtain a construction/permanent loan at the initial application. This action insures that you settle on both loans simultaneously and hence that you pay settlement costs only once. For a construction loan, you may want to consider borrowing more money than you actually figure that you will need. You will obtain the money by drawing on it as you need it to complete construction. If you don't need all the money, you do not have to draw it. On the other hand, if you do need it, the money is there, thus relieving you of worries as to where the money is coming from to finish construction.

Shop for the best loan.To get the best loan, shop and negotiate. To lending institutions, loans are commodities which follow the economic law of supply and demand. There is also competition among lending institutions, and since there is, the shopper can keep searching and negotiating until he finds the best bargain. One should not hesitate to ask for reduction in rates, for lower numbers of points, for more attractive terms at prescribed interest rates, and, elimination of the one-percent loan origination fee. The latter should be a strong point on the part of a borrower who has a strong financial statement. Since I have already had the experience of a bank reneging on the rate that it had agreed to freeze, I do not intend to let it happen again. The next time I apply for a loan, I intend to ask at the inception that the lending institution notify me in writing that once it freezes the interest rate, it cannot renege, regardless of any action that might be taken by the Federal Reserve Board or by any other outside influence. After all, the institution would not have honored my reneging on a rate and attributing the reason to an outside action.

Provide accurate and timely information. To put yourself in a strong bargaining position, it is vitally important that you provide the bank with timely information on everything that they request, and that it be done truthfully and straight forward. It is also important that you have all your personal affairs in order, so that you portray the image of one who is mentally stable in every respect, has no overdue and unpaid debts of longer than thirty days duration, and is financially trustworthy. If you can portray this picture, you are encouraged to shop until you find the best offer, even if it means calling the mortgage loan officer in every financial institution in the yellow pages of the telephone directory.

It has been pointed out elsewhere in this book that banks and lending institutions tend to enjoy special privileges under the law; hence, one must realize from the inception that although one can negotiate with a lending institution in attempting to obtain the most favorable terms, those terms are never locked in concrete until all documents are signed and the money has actually changed hands. History is replete with cases of individuals who have thought that they have locked-in a given interest rate only to learn that, for one reason or another, the lending institution has reneged on its verbal statements. In such cases, I have never heard of a case where the loan applicant has been successful in so much as having his complaint tried in court. It is therefore important for a mortgage loan applicant to comply as astutely and as expeditiously as possible with the institution's request for information and identified documentation and to close the loan as soon as the bank is ready. There have been several times in recent years when the exigencies of the money markets and accompanying action of the Federal Reserve Board have caused interest rates to soar or plunge overnight. When the former occurs, it can wreak havoc on the pocketbooks of mortgage loan applicants who have planned on monthly mortgage payments within a narrow range. It is therefore prudent to nail down suitable terms at the moment they become available. When my wife and I failed to do so, an overnight rise in the mortgage rate caused the lender to cancel our rate freeze, thereby forcing us to accept a short term adjustable rate mortgage. Two and one-half years later, we are still awaiting the development of favorable terms.

The three main general types of loans: When time comes for you to obtain your long term measure, there are a number of options that the lending institution may offer you, so numerous in fact that it is impractical to mention more than a few here. First, there are three general types of loans, a Federal Housing Administration (FHA) loan, a Veterans Administration (VA) loan, and the Conventional loan. All three are generally offered by almost every lending institution. When you acquire an FHA loan, the federal government agrees to guarantee payment of a high percentage of the loan, in return for your making a prescribed down payment and paying a mortgage insurance premium of approximately 3.7%.If you make a down payment of a minimum of twenty percent of the loan value, you will not have to pay the mortgage insurance premium. The VA loan is somewhat similar to the FHA loan except, one has to be a veteran of the armed forces to become entitled to it. Once used, it cannot be used a second time until the initial loan has been fully paid. Under a VA loan, it is possible for a veteran to buy a home with no down payment, with the federal government guaranteeing payment of the loan. VA loans can generally be obtained at one-quarter to one-half percentage points lower interest than conventional loans,because the Veterans Administration sets limits that banks may charge for interest on VA loans. This limit is sometimes circumvented by lending institutions charging points for processing VA loans. Another advantage of FHA and VA loans is that they can be assumed by others who may buy the property in later years, thereby providing the potential benefit of making the loan financially desirable if the prevailing interest rates happen to soar during coming months or years. The conventional loan is offered by each lending institution on its own terms which are controlled only by the law of supply and demand in competition with other lending institutions. Seldom are conventional loans assumable.

Other types of mortgages: Either a VA, FHA, or Conventional loan may be acquired for a term of 15, 25, or 30 years, or other lengths of time, depending upon what the borrower wants to work out with the lending institution. Other types of mortgages that may be acquired under conventional mortgages are adjustable rate mortgages (ARMs) whereby the borrower agrees with the bank for it to adjust the interest rate periodically to bring it in line with current mortgage rates. The lending contract spells out the intervals that must elapse between adjustments and sets a limit that may be applied to each adjustment.

The determination of interest rates: Current interest rates are generally determined by referring to the average rates on government treasury bills or instruments of the Federal Home Loan Bank Board.

ARMs are popular when interest rates are so high as to make the borrowing of money extraordinarily expensive. In subscribing to an ARM, the borrower is counting on the interest rate going down in subsequent years. By law, the bank must apply the same criteria to adjust rates downward if the criterion goes down during the period between adjustments. There are a number of other types of mortgages. One of them is the graduated-payment plan, whereby the monthly payments are small for the first few years, gradually escalating each year until a few years later when they become fixed. If one feels sure that his income will rise in future years, the graduated payment plan may be worth considering.

What the bank will request from you: When you apply for a construction loan or a long term mortgage, the lending institution will ask for a list of your assets and liabilities, some type of financial statement, a statement from your employer concerning your salary, detailed information concerning the property you are buying, and a fee to pay for appraisals or other administrative requirements. If you are applying for a construction loan, the bank will ask for all the above and upon approval of your loan, will set aside an amount that you may draw against as you progress in the construction effort. You will have to attain specific milestones for each withdrawal of funds that you obtain.

NINE: DOING MUCH OF YOUR OWN CONSTRUCTION

Depending upon the skills possessed by individuals contracting for the building of a house, there are many areas wherein considerable savings can be realized by the individual doing a portion of the work that would otherwise be done by subcontractors. Even without substantive talent or experience there are areas where the expenditure of a little physical energy will improve the overall quality of the finished product. Most of the work going into the construction of a house is much easier to do than it appears.

Possible areas for the client to apply self-help: Among these are painting the interior and exterior of the house, wall papering, installing chair railing and ceiling molding,constructing window valances and cornices, final cleaning of the house inside and outside,final raking and seeding of the lawn, laying of straw to prevent erosion of seeded areas, landscaping, finishing complete basement areas, building of decks,self fabrication of window treatment material, caulking, self installing intercom and burglar alarm system, installation of ceiling fans and selected lighting fixtures, self installation of inter vacuum systems, and other tasks that may be within the repetoire of talents of the party contracting for the construction of the house. Many people have latent talents that they have never had a chance to apply. Many have discovered to their own amazement that they have been able to build cabinets and to lay subflooring. Others have found that they haven't been able to do all the work themselves, but by calling for only a little outside help at near minimum wages, they have been able to do a creditable job.

Most general contractors are reluctant to agree to client participation: One must understand that many builders will be extremely reluctant to grant clients the prerogative of self-help, for it is in many of the areas mentioned that the builder makes his most lucrative profits. For instance, some builders purchase mistinted paint from vendors at only a fraction of the cost of properly tinted paint; and then apply it to the building as the prime and second coat. The paint work may appear to be of professional quality, but it will pose insurmountable problems a year or two later, when for one reason or another, the house owner tries to match it. If the builder declines to permit the owner to perform selected work during the construction process, the owner should look for another builder. Many of them will gladly grant the owner the self-help prerogative, and some will even provide advice and a moderate amount of assistance. Generally, it is in the area of trimming and finishing that the work can be effectively accomplished by the client.

Look for a builder who will agree to the self-help prerogative: There is a middle road that can be taken by one contracting for the building of his home, when he really would like to do his own subcontracting. That road is for him to find a builder who only recently has acquired his builder's license and is a little hungry for a contract. Negotiate a contract with him whereby he does most of the work but leaves selected elements for you to perform. There are many general contractors who will subscribe to such an arrangement.

The important consideration is that most people can do a lot more of the tasks associated with house building than they realize. My wife and I performed all the tasks enumerated above plus a number of carpentry tasks, such as hanging all the doors and installing all the door hardware. We even installed all the ceiling rafters in one of the upstairs bedrooms,and by diligently watching the drywall crew perform their work, I was able to do a creditable job of wall installation throughout the eighteen hundred square foot basement.

Consider attending one or more of the short home-building schools: If you have the time and are willing to apply yourself to a selected number of housebuilding tasks, you can do them. They are not that difficult. If after reading this bit of encouragement, you still have doubts about your ability to participate in the building of your own home,there is one further alternative. There are a number of do-it-yourself homebuilding schools proliferating now. They offer courses in just about every type task associated with home building, including everything from wiring, painting and plumbing to carpentry even budgeting. The courses range in length from one to three weeks at costs per course ranging from $350 to

$500. Today, labor costs on a house are from 50% to 60% of total costs, so anything that you can do yourself will save you from 50% to 60% of the costs of that particular task.

You may want the general contractor to complete only a pre-scribed portion of the building project: If you can afford to wait for several months or even a year or two before you have a home that is 100% complete, the sky is the limit as to the amount of self help that you can put into your home. There is no law on the books that says that you have to contract for a completed house. You can contract for any portion of a house. Many people have contracted for only the shell of a house to be completed, and then they have finished the work on the house themselves. With today's high construction costs, a good case can be made for having the general contractor to install the foundation, complete the framing, roofing and boxing and leave any portion desired for the client to finish at his own leisure and at his own expense. Many people have gone a bit further and have had one or two rooms completed in their entirety, leaving several rooms to be completed with their own labor while living in the house.

There are at least a hundred variations of how much can be done by a contractor and how much is to be done by the owner. In addition there is always another alternative of having the general contractor complete the house to a certain point, at which time the owner assumes responsibility for completing the house through a combination of self help and managing his own subcontracting. Another possibility in the do-it-yourself category is to hire individual workmen. You may know college students or people interested in earning a little additional money. If you know them well enough to be certain that they can do the work and respond to instructions, you may want to hire them as individual workmen. The advantage would be that you could hire them for less than half the costs of subcontractors. The difficulty in this approach is that if the work is of any appreciable duration, say more than two weeks for any one person, you would be required to pay social security, workmen compensation, and unemployment insurance, along with the necessity of keeping numerous records and filing reports.

Any combination of work being done by contractor, self help and owner managed subcontracting can generate considerable savings for the owner. It depends entirely upon how much free time the owner may have and how long he is willing to wait until his house is completed. My wife and I left one of the upstairs bedrooms unfinished, so we could complete the room at less expense. One cannot detect a difference in quality of workmanship between the room we completed and other rooms in the house which were completed by the general contractor.

In most states, one cannot build a home for another person or for resale unless he has a general contractors license. However, the law does not restrict one by demanding that he be licensed before he is permitted to build a home for himself. When one does build for himself, however, he is required to follow the same building codes that a general contractor is required to follow. He must obtain a permit and must call for periodic inspections as he progresses through the various construction phases. This requirement can sometimes be a blessing in disguise by creating a source of advice for one who has encountered a problem for which he has not been able to find a ready answer.

TEN: THE BUILDING INSPECTOR

All of the fifty states have building codes which have been approved and enacted into law by the legislative process. There are codes for various types of commercial and industrial structures, but there is a separate code book for residential dwellings. Formulation of the codes have been for a wide variety of purposes, but among the most important are the motives of protecting the public from fire, safety and health hazards. There is the additional motive of assuring order and esthetics in development throughout the state with an eye toward precluding needless destruction or spoilage of the natural environment.

The following is quoted from a pamphlet entitled, "NORTH CAROLINA UNIFORM RESIDENTIAL BUILDING CODE", prepared by North Carolina Building Inspectors' Association - 1968 Edition with Amendments thru December 10, 1985:

ARTICLE I. GENERAL

SECTION I. Citation of code; to what structures applicable.

The following provisions shall constitute and be known as the uniform residential building code and shall be cited as such and provides for matters relating to construction, alteration, repair, or removal of buildings or structures, erected or to be erected in the state. The provisions of this code shall apply only to residence buildings, duplexes or structures hereafter erected, and to any alterations to existing buildings but does not apply to apartment or multifamily houses constructed, altered, repaired or used as a residence for three or more families.

SECTION 2. Application, plans and permits.

Before the erection, construction or alteration of any building or structure, or part of same, there shall be submitted to the Building Inspector, by the owner or authorized agent, an application on appropriate blanks to be furnished by the Building Inspector, containing a detailed statement of the specifications, and accompanied by a full and complete copy of all necessary plans of such proposed work. Each application for a building permit shall be accompanied by a plat, drawn to scale, showing accurate dimensions of the lot to be built upon, accurate dimensions of the building to be erected and its location on the lot. If it shall appear to the Building Inspector that the provisions of this code and the State building laws have been complied with, and all requirements of fees has been paid, he will then issue the building permit. A copy of the plans as approved by the Building Inspector shall be kept at the building during the progress of the work and shall be open to inspection by the Building Inspector. Plans and specifications submitted to the Building Inspector shall be kept in files in his office or returned to the owner. It shall be within the discretion of the Building Inspector to issue permits for minor construction work without plans and specifications.

The above two articles define the code and how it is applied. The code book goes further to state that the schedule for building permit fees are to be regulated by each city, town, or county, also each contractor's bond or liability insurance is to be regulated by each city, town or county. The code authorizes the local Building Inspector "at all reasonable hours, for the purpose of examination, to enter into and upon all buildings and premises in their jurisdiction." Further, the code says that as the building progresses, the inspector shall make as many inspections as may be necessary to satisfy him that the building is being constructed according to the provisions of the law.

A good building inspector is the best friend that a home-buyer can have: The Building Inspector performs a valuable service for one contracting for the building of his home, as well as for one who does his own subcontracting. The building profession has come to depend heavily upon the building inspector, sometimes too heavily. Many general contractors don't even bother to inspect the finished work of subcontractors who have performed specific tasks. They depend upon the subcontractors to call for an inspection when they have completed their work. The office of the Building Inspector responds to the request within a matter of twenty-four hours and performs the inspection for a fee of thirty dollars(in the area where my house was built). If he approves the work that the subcontractor has performed, the subcontractor provides a copy of the approved action to the General Contractor as evidence of successful performance of his work and in support of his request to be paid. If the Building Inspector disapproves the work of the subcontractor, he tells him what he needs to do to make the work acceptable and leaves. The subcontractor can then request another inspection when he thinks that he has corrected the faulty work to the satisfaction of the Building Inspector. The second inspection requires payment of a fee of $13.00. Most General Contractors require the subcontractors to pay the inspection fees.

Problems that are beyond the scope of the building inspector: Occasionally, there will arise a complex situation during the course of construction that will demand a degree of advanced technical or engineering knowledge beyond the scope of the Building Inspector. Such a question may arise as a result of the need to determine the stress or the bearing of specific structural members of a building, the computation of which requires a person with advanced engineering training. In such cases, the law requires that the question be answered and certified correct by a licensed structural engineer, whose certificate must be accepted by the Building Inspector.

If you are subcontracting your own home, you will become acquainted with the Building Inspector who inspects the work of each of your subcontractors. The Building Inspector teams are good people to know. Chances are that you will require the subcontractor to pay the inspection fee,but you will certainly be interested in learning what the Building Inspector has found wrong, and you will undoubtedly have questions which you will want to ask. For the fees that are paid, you will get assurance of a safe and reliable home.

With the rapid advance of technology, there is an increasing tendency for materials to be developed which improve the state of the building art, especially in materials which provide for greater structural strength and in more effective fire retardants. The Building Inspector may at times be called upon to determine whether such an item is an adequate substitute for the item specified by the code. The code states that in interpreting the code, the decision of the Building Inspector shall be final.

Before making his final decision,however, he is expected to confer with engineers who are prominent in the building industry, or any other source where current information may be available to assist him in his determination. Once a decision is made however, it may be changed only by taking the decision to the local board of appeals, the State Building Code Council, or to the courts. The Building Inspector has the authority to permit the use of materials or methods of construction not specifically set forth within the code, provided any such alternate materials or methods of construction is proved to the satisfaction of the Building Inspector to be at least equivalent of the requirements prescribed by the code for safety, strength, quality and effectiveness including fire resistance.

During the building of our house, my wife and I had two occasions when the decision of the Building Inspector had to be complemented by the technical knowledge of a licensed structural engineer. The inept builder whom we had been forced to fire had changed the design of the great room without any knowledge of how he was to span the twenty foot spacing between walls. The Building Inspector had no ready answers except the possible use of a twenty two foot flitch plate sandwiched between 2"x10"x20'timbers. There was doubt as to the effectiveness of the flitch plate because of the tendency of metal to expand and contract with the changes in temperature. We called upon a licensed structural engineer who acquainted us with a late development of a beam consisting of compressed plys,with strength which actually exceeded that of the flitch plate. On the other occasion, the structural engineer was also the answer to our problem. The general contractor had installed a cat walk in an area where posts had been designed by the architect. The structural engineer showed us how to use plywood nailed to the sides of short upright stud sections to make a knee wall which solved the dilemma. The Building Inspector was happy to obtain the structural engineer's recommendations on both occasions.

ELEVEN: TWENTY SIX TIPS FOR SAVING MONEY

Whether you do your own subcontracting, or contract with a professional builder, there are myriad ways that you can reduce the overall cost of the house without degrading the quality of the final product. In Chapter 3, we listed a number of ways that a general contractor uses some of these avenues to increase his own profit margin. However, you are interested in saving money for yourself, not the builder. With this personal objective in mind, you may be willing to accept some of the features to which you would object if done by the builder. In the discussion below, there are listed twenty-six tips that you may employ to reduce construction costs.

1. Obtain contractor discounts. Building supply companies have at least two prices for their material, a price for the general public and a discount price which is quoted to contractors. These discounts are provided in a number of ways. One building supply company that I know provides a contractor a 5% discount on every invoice, and provides the contractor an additional 2% discount if he pays his bill by the tenth day of the month following purchase. Another firm provides a straight 6% discount, while another has discount rates varying from 2% to 25%, depending upon the class of material purchased. General Contractors seldom advise their clients that these discounts are being obtained and quote the retail price of the products in their estimates of costs of materials. If you intend to do your own subcontracting, or to do any of the work when you contract with a builder, you will save a considerable amount of money by visiting the credit managers of several building supply companies and inquiring as to the discount rates that may be available to you if you set up an account with them. When I was forced to resort to doing my own subcontracting, I set up accounts with three vendors and thereby was able to obtain the best prices available on items that I purchased.

2. Use finger joint molding. The trend in today's housing decor is toward spaciousness and naturally lighted interiors. Much of the stained woodwork of the past is giving way to trim in varying shades of white. If you intend for your baseboards, chair and crown molding, and window and door facings to be painted, don't waste your money by buying the most expensive molding. If the item is to be painted rather than stained, the finger joint molding will serve your purposes equally as well as the much more expensive hardwood molding which would be necessary if you were to stain it. If there are selected places in your home that you still desire some stained wood work, quality effect can still be achieved by purchasing the standard grade of molding and treating it with a coat of wood conditioner sold in building supply activities or hardware stores. The conditioner has the effect of penetrating the wood to seal the most porous areas, thereby permitting the application of stain and polyurethane coats without leaving the mottled appearance that is prevalent in so many of today's new homes.

3. Install your own crown molding and chair molding. If one owns a miter saw, chair molding and crown molding can be installed without difficulty. If you don't own a miter saw, consider purchasing one of the economy models, currently available at a cost of about $90. Buy your own molding from one of the discount suppliers with whom you have established an account, and you will need nothing more than a hammer, a package of small finishing nails, a $1.50 caulking gun, and a $1 tube of white caulk. With those few items and a five foot wooden ladder (which you will need anyway if you are to own a home) you can save approximately $300 if you plan to install molding in as many as four rooms. You will save more if you do more than four rooms.

4. Consider doing most of your interior caulking. Before the interior of a house is painted, it is necessary to caulk along the top of all the baseboards and around the facing of all windows and doors. This is a job that you can do with only a caulking gun and a few tubes of caulk. If you pay to have it done at $10 to $20 per hour labor costs, it will cost you approximately $500 for a 2500 square foot house.

5. Consider the installation of plastic doors, wherever practical, instead of the more expensive wooden doors. The average wooden door costs about $140; the average plastic door about $35. It may be that you will prefer wooden doors in the prime living area, master bedroom, kitchen area and some other areas that are in conspicuous places. However, there will be other places such as closets, upstairs bedrooms, basement, and stairwell that you may be willing to settle for the plastic doors, keeping in mind that todays plastic doors are difficult to distinguish from wooden doors unless one makes very close examination. If you install plastic doors in only ten locations, you will save approximately $1,050.

6. Consider installing all your door hardware. Quality door knob assemblies and hinges for a door costs about $20. If you buy at discount prices, you can get the hardware for about $18, thereby saving about $2 per door on the cost of the hardware. If you pay for the installation, you can expect to pay at least $12 per door in labor costs. Thus, if you install the hardware on twenty-five doors, you will save approximately $350 in labor costs. Tools needed will be an electric drill, a key hole saw attachment, a screw driver, and a chisel. You may already have most of these items in your tool chest. If not, you should buy them for future home maintenance and repair. The total cost of all, including the drill is about $50.

7. Consider using factory reject plywood for subflooring. Factory reject plywood is generally as strong and durable as that which has passed inspection. Often it has been weather damaged, or has an excessive number of knots. These imperfections detract from its appearance, but not necessarily from its utility. The factory reject plywood is normally available at less than 50% of plywood retail costs. As it is to be used for subflooring, it will be covered and will function only for the purpose of providing support for the main floor. The cost of factory approved plywood is about $2.40 per square foot. Thus the cost of plywood subflooring for a 2500 square foot house is approximately $6,000. If by using factory reject material, you can save 50% of this cost, your savings will be $3,000. If you recall, we mentioned plywood subflooring in Chapter 3 as one of the ploys sometimes used by builders to increase their profit margins. The difference is that if you are doing the job yourself, you will not use a piece that has been damaged to the extent that it will result in an inherent weakness in the subflooring.

8. Consider painting or staining the siding before it is installed. Most houses, even those constructed with brick veneer, have cedar or fir siding on portions of the house. The specifications will call for this siding to be stained or painted. You can save a considerable amount of money, depending upon the total square footage of siding to be installed, by painting or staining it before installation. This method has at least two advantages. First, the paint or stain is easier to apply on a flat surface, and secondly, it adheres or soaks into the wood more readily than it would if applied after the siding has been installed. It is easy to lay the 4'x9' pieces of siding on two saw horses and to apply the paint or stain with a roller, using a brush only in those areas where there are grooves that have been designed into the siding. My wife and I stained approximately 100 sheets (3600 square feet) of such siding while our house was under construction, and thereby eliminated the need for hiring exterior painters after its installation. If we had hired the staining done, the costs would have been approximately $3,400.

9. Finish your garage and basement. The specifications for most houses don't require finished garages or basements, yet many people add finishing as a requirement for the general contractor, who of course charges sizeable extra fees for the work. Don't engage a contractor for this type of work. Do it yourself. The garage is fairly easy to finish. The studs have already been installed on 16" centers and the space between the studs will already have insulation installed. All that you will have to do will be to purchase the necessary number of sheets of 4'x8'dry wall at about $4.75 per sheet, buy a package of dry wall nails and proceed with nailing the sheets side by side. At corners where you need less than a full sheet, you will use a carpenter knife to score a straight cut down the sheet on both sides to the width that you have measured and then snap the wallboard to break it at the desired width. Professionals normally use 4'x12' wall board and install it laterally instead of vertically, but 4'x12' dry wall is too heavy to be handled by only two people, so you are advised to use the 4'x8' variety and install it vertically or horizontally according to the easiest approach dictated by the size of the garage. The ceiling wall board generally can also be installed on existing ceiling rafters, but for one or two people to install it requires the fabrication of a temporary scaffolding. You may want to ask a local wall board mechanic to advise you on the scaffold. Nailing the wall board to existing studs and rafters is followed by the application of wall board tape over all cracks and joints, glued on with wall board paste available at building supply companies. You may want to hire a wall board mechanic to seal the seams, but you can become adept at the sealing process after an hour or two of mistakes. You need to know that the process is done in two, sometimes three, increments letting the earlier application dry for two or three days before applying the next paste. If you have purchased the miter saw mentioned above, you will find that the installation of the baseboard is easy.

When I asked for quotations for finishing my garage, the total fee was $1,100. You can save this amount by doing it yourself. Finishing your basement will be a little more difficult than finishing the garage, but it can be done by a beginner if he is prepared to take his time. Normally, a basement will have walls that are of concrete block. There may be no previously installed studs. You may need to install the studs yourself. This is easy to do with nothing more than a saw, a hammer, a measuring tape and a bucket of nails. In my case, I was able to find nearly all the lumber I needed by picking up discarded 2"x4"pieces from the scrap pile that careless sub contractors had cast aside. Inasmuch as the stud sections were not to be load bearing,I spaced them on 24 inch centers instead of the customary 16 inches. I installed insulation between studs and proceeded in the same manner as we specified above for the garage. After installing the wall board, I installed a suspended ceiling, something that I had never done before, but which I did successfully by following directions encompassed in each package of ceiling tile that I bought.I bought damaged doors from a local building supply company at a cost of $5 each and installed five doors at appropriate places in the basement. I installed baseboard and floor molding all the way around the 1800 square foot basement,and then hired a subcontractor for carpeting. The result was a beautiful basement that, in appearance was as beautiful as the upstairs. The estimated cost of what a subcontractor would have charged for the entire basement is about $20,000 compared to the $4,300 that it cost me to finish the job myself.

10. Consider installing your own intercom system. My wife and I installed our intercom system after our house was completed, simply by following directions included with the material. By not doing it until after the house was completed, it was far more difficult than it would have been if we had done it before the insulation and dry wall were installed. We were successful on our first try. Estimated savings was $750.

11. Consider doing all your own interior painting. When my wife and I asked for bids for the interior painting of our house, the bids ranged from $4,900 to $5,500 for the upstairs 3375 square feet. The bid did not include the 1800 square foot basement. We elected to do the work ourselves. We finished applying the first coat in about four days, then waited about two weeks before applying the second coat. We finished the second coat application in three days. Of course, we worked ten to twelve hours each day. The results were every bit as good if not better than if we had hired subcontractors. Our total effort cost us about twenty gallons of paint and a few brushes and roller sleeves. We saved approximately $5,000.

12. Consider installing your own closet shelving. You don't have to be an expert carpenter to install closet shelving and clothing rods. You need nothing more than a tri-square, adequate amount of one inch by twelve inch pine or spruce lumber, and a few nails. Take a look at the shelving in any model home and duplicate it for your own home. By doing the work yourself, you will save an average of eighty dollars per closet. If you have ten closets, your savings will be approximately $800.

13. Consider installing at least one or two bath rooms. Depending upon the number of bath rooms that you intend to have, you can save an appreciable amount of money by assuming responsibility for the completion of one or more of them. When our house was constructed, we had a bath room "roughed-in" in the basement, so that we could easily install the necessary fixtures at a future date. When we finally decided to proceed, we asked for quotations from subcontractors on the cost of installing a toilet bowl, a sink, and a sewer pump. The quotations ranged from $1,300 to $1,500. We elected to buy all the equipment ourselves at contractor discount prices and purchased the necessary fixtures and appliances for a total of $310. We then had two plumbers working at a rate of thirty dollars per hour to install the system. The total bill for the installation was $162. The bathroom installation cost us a total of $472, thereby saving us $928. When I asked the reason for the disparity, a plumber told me that any system installed by him as a subcontractor incurred an obligation on his part to be responsible for the system for at least a year, maybe longer; that inasmuch as I had assumed responsibility, he would always work as required for thirty dollars per hour.

14. Consider building your own deck. Like the closet shelving, a deck can be built by a novice carpenter. It generally consists of little more than several four inch by four inch timbers encased in concrete footings. The deck is anchored to the house at one side with heavy bolts. It is rough carpentry consisting of a 5/4 inch deck nailed to several 2"x10" stringers spaced on sixteen inch centers. For my house, I asked for quotations for a twelve foot by thirty-two foot deck. The lowest quotation was $2,000. I built it for a total cost of $600.

15. Consider doing your own lawn seeding. For a one acre lawn, subcontractors quote prices ranging from $1,100 to $1,300. This includes raking and disposing of rocks, sowing the seed, and putting straw down to prevent wind and water erosion while the seed are germinating. My wife and I elected to do it ourselves. We bought twenty sacks of fescue seed at $20 per sack and 26 bales of straw at $3.50 per bale, at a total cost of $491. We were told by a friend that complete raking for rocks was unnecessary, because rocks were prevalent throughout the soil and would surface after each rain until the grass began to cover them, so we gathered only the large rocks, leaving the smaller ones. We found later that our friend had been correct. When the grass emerged, we had no rock problem. Nevertheless, our seeding effort saved us a minimum of $609.

16. Consider doing your own landscaping. You can save a great deal by buying and planting your own shrubbery. We chose not to hire subcontractors and to delay landscaping until after we moved into the house. Subcontractors had quoted prices for shrubs that we knew that we could buy for half the price. We heard that we could get far better prices if we travelled about twenty-five miles from the city to a small nursery that wasn't receiving large orders daily like those in the city. This report proved correct. We were able to buy English Boxwoods for $15 per plant, whereas local nurseries had been receiving as high as $35 per plant. We bought all our shrubbery at the out of town nursery at prices that averaged about forty percent less than the city nurseries. Since we planted the shrubs ourselves, we saved about one-half the cost of what we would have paid a subcontractor. As initial landscaping quotations were in the area of $2,500, we saved about $1,250.

17. Buy trees at the end of the season. We wanted a number of Bradford Pear trees for our yard. We were advised by a local vendor that if we waited until almost the time of the first frost that these trees would be on sale for 50% to 60% of their regular cost. This also proved correct. We were able to acquire twelve Bradford Pear trees at a cost of $12 each, for a total cost of $144. If we had bought them earlier in the year, they would have cost us $240. Paying to have them planted would have cost an additional $10 per tree. We saved $216.

18. Buy all your plumbing fixtures. The rule of thumb for subcontractors in the plumbing business is to plumb a house for a general contractor for a price that is computed on the basis of $260 per fixture; i.e., if a house is to have 20 fixtures, the bid for plumbing the house is 20 times $260, or $5,200. This price includes the fixtures to be installed such as toilet bowls, sinks, hot water heater, laundry tub, outside faucets, and any other type fixture that the house may have. In providing these fixtures the plumber installs standard builder models in every spot, unless you express your desire for something more elaborate. In either case, you lose. If you accept the builders model, he buys it at contractor discount even though the $260 fee quoted was for retail price. If you express a desire for something more elaborate, the subcontractor tacks on an additional fee on top of the $260 to make up the difference between the cost of the builder's model and the one you have selected. You can partially counter this ploy by buying all the fixtures yourself and asking him to deduct the cost of the builder's model from his fee. In this manner you will be the recipient of the 15% discount on the plumbing fixtures. The plumbing subcontractor won't necessarily like it, but he will accept it rather than attempt to explain how he had proposed to make the added profit. We purchased about $2,200 worth of fixtures, receiving a discount of $330. That was a savings to us. If we had done it the subcontractor's way, he would have gained the $330.

19. Consider 3/8 inch wallboard for all interior walls that have wallboard on both sides; i.e., as opposed to those walls that constitute the perimeter of the house. One-half inch is generally used indiscriminately, but when insulation capacity is not needed as in the case of walls completely inside the house, you can save money by using 3/8 inch wallboard. The costs of 3/8" wallboard is about five cents per square foot less than 1/2" wallboard, so if you have 3000 square feet of interior walls in your house, you can save about $150 by using 3/8" wallboard.

20. Consider installing ready made cabinets and counter tops, rather than having them custom made. Ready made cabinetry can be purchased from many building supply activities in a wide array of hardwoods and in many different sizes. They can be bought for half the costs of custom made cabinetry. If you buy unfinished ones, you can finish them yourself, using nothing more than a $25 palm sander. #100 sandpaper, and a gallon or two of polyurethane. After finishing them you can have your builder install them, if you feel it is necessary, but you can do it yourself. As many homeowners pay as much as $12,000 to $18,000 for custom made cabinetry throughout the house, you can save as much as $5,000 to $9,000 by doing it yourself.

21. Consider wall to wall carpeting in some of the rooms where you may have intended to install hardwood flooring. You can do it for approximately 1/3 the cost. You can still have selected rooms with hardwood flooring, but if you substitute carpeting in only four rooms, you can save as much as $2,500.

22. If you have more than two bath rooms, consider economy type plumbing fixtures in the third and fourth bathrooms. Toilet fixtures tend to vary in price by as much as 400 %, depending upon the style that you select. Yet, the difference is sometimes hard to detect. You can go with the more elaborate fixtures in your main bath and guest bath if you like, but you may want to consider saving about $400 per bath room with less expensive fixtures.

23. Consider installing the black economy type insulation board for the outside sheathing of your house, eliminating as much as possible of the foil covered sheathing. You won't need the extra strength if you are using fir or cedar siding, or any other type of wooden siding, and the insulation quality will still be there. On a 2500 square foot house, you can save as much as $1,000.

24. Enhance the style and increase the value of your home by installing ceramic tile in the bathrooms and kitchen. This is not an option which returns an immediate savings. The economic benefit comes when you sell your home several years later. Ceramic tile is more expensive than carpeting and costs about the same as hardwood flooring, but the benefits are its durability, enhanced home value, and maintainability. Tiling a kitchen or bathroom costs between $6.50 to $7.50 per square foot for total costs of labor and materials. If you're lucky enough to do it at a time when there is a slow down in construction in the general area, you may be able to get it done for as little as $5.50 per foot. Once installed, you can expect it to last a lifetime,whereas you would have to replace carpeting about every five or six years. You can save at least $4 per square foot if you do it yourself; however, you are cautioned that the work takes a great deal of patience and accuracy of alignment. You can acquire pamphlets on how to do it at most building supply activities, and if you follow instructions precisely, you can do a creditable job.

25. Consider eliminating some of the more expensive frills that will not pay you back when you sell the house. There are a number of added features in all middle and upper income homes that are designed to provide a degree of extra comfort, increase the building's curb appeal, or simply to please the fancy of some member of the family. Look carefully at these items to see if you are really willing to pay what they will cost you during the time that you own the house. Many added features are worth the time and expense going into them. Many are not. Listed below are the features that go the longest way toward amortizing themselves when the building is sold.

26. Specify in your contract that you will provide all kitchen appliances, all light fixtures, the intercom system, and the carpeting. Then start shopping for the best prices. If you go to the various building suppliers and explain what you are doing, you can get the same discount that a general contractor can get. In most cases you can obtain better prices than the builder can obtain, because you are shopping to save your own money, not someone else's. By intensive shopping and negotiating, you may save as much as $400 on a refrigerator. When I built my house, I found that I was able to beat the general contractor's best price on the same carpeting by nearly $600. Busy general contractors do not have the time or the patience to shop vigorously for the best price. If you don't enter the picture, they will obtain for you the item at routinely discounted price, a cost which may exceed merchandise at discount stores by as much as 25%.

Table 11-1. Added features that pay you back

Added feature	Approximate Cost	Return of cost
Wood deck or porch	$3,000 to $9,000	40 to 75%
Additional bathroom	$3,000 to $12,000	60 to 100%
Two-car garage	$15,000 to $30,000	60 to 100%
Central air	$3,000 to $8,000	40 to 100%
Added kitchen space	$50 to $100/sq. ft.	40 to 100%
		unless you make it too large
Workshop,exercise or laundry room in Basement	$2,000 to $8,000	75 to 100%
Energy-efficient fireplace	$1,000 to $5,000	60 to 85%
Solar greenhouse	$8,000 to $35,000	50 to 80%
		upper income homes only

Table 11-2. Features that do not pay you back*

Added feature	Approximate Cost	Return of cost
Swimming pool	$5,000 to $30,000	0 to 5%
		In some cases value decreases
Tennis court	$3,000 to $7,500	0 to 15%
Intercom system	$2,200 to $5,000	10% to 25%
		More for very expensive homes
Basement rec room	$3,000 to $10,000	40% to 50%
Central vac system	$4,000 to $8,500	20% to 30%
		More for expensive homes
Gazebos,dog houses, tree houses, cabanas	$1,000 to $10,000	Maybe 1% to 2%
Elaborate landscaping	$3,000 to 18,000	20% to 30%
Screening and fencing	$1,000 to $15,000	40% to 60%
		for first $5000. Little after.
Solar heat	$10,000 to $35,000	10% to 25%

*If you are an avid gardener or sports enthusiast you may reap full value for the above items in terms of your own or your family's actual enjoyment. Otherwise, each of the above is a money loser when the time comes for resale. You may want to consider changing the plans if they contain some of the above items which will not amortize themselves in terms of your personal satisfaction.

TWELVE: AWAKING FROM THE NIGHTMARE

The narrative below portrays the experiences we had in recovering from the nightmare evoked by our failure to do our homework before contracting for our home. With our resolute determination to recover, we began to get a firm grasp on the details of the problems confronting us and to do whatever became necessary to solve those problems.

Assessing the damage. We employed two of the more reputable builders in the area to evaluate the house and to provide written reports on remedial construction effort that would be necessary. Each builder conducted the inspection and provided a report for a fee of $300. Each report stated that corrective action preparatory to proceeding with construction would cost approximately $14,000.

Engaging Attorney No. 2. By now, we recognized that instead of the $206,000 which Jaker had estimated as the cost of the building, it would be approximately $250,000. If you recall, I told you at the end of Chapter 2, that an attorney had recommended another attorney to me, that the name of the new attorney was Haywood Coward. The lawyer recommending him had said that Haywood was like a bulldog, that he would grab on to an issue and never let go. When my wife and I went to Haywood's office, he asked us to convey the entire story to him. I said, "I have typed every item in chronological order in narrative fashion. I suggest you read it now. The document is yours to keep so you won't have to take notes on what I relate to you". I asked him to pay careful attention to the documented indiscretions of the bank which had loaned Jaker too much money on our house in disregard for the welfare of my wife and me as the third party. I asked him also to examine the legal data search which I had included with the documentation which I was providing him. I said, "Tell me the fee that you charge for your services". He replied, "I charge $105 per hour, but that is not important. If I get your money back, it will be well worth it". He spent about twenty minutes reading the documentation and said, "I am not sure that you have a case against the bank. We would have to prove that the bank knew that third party interest was involved. After we get into the matter, we may be able to force Jaker and his wife to testify in detail regarding their dealings with the bank and, if so, we may have a suit against the bank". He suggested, "Right now, I think we should go after Jaker. He may not have any money, but if we scare him badly enough, he will obtain the money in some manner similar to the manner in which he obtained yours". I said, "The best way to make Jaker come up with the money is to convince him that he and his wife are going to be criminally prosecuted. If you review the case, you will see that Jaker knew exactly what he was doing. We can prove that he had criminal intent to misuse our money". Haywood replied, "It is now against the law for an attorney to threaten an adversary with imprisonment, but there is nothing wrong with my telling him that you and your wife are interested in making him and his wife the object of criminal prosecution".

A few days later, I received a statement from Haywood billing me for $685 for time which he claimed that he had spent reading the material that I had witnessed him read in his office in only twenty minutes at the time that I had given him the documentation. Jaker immediately retained his own legal counsel. After procrastinating for several months and having his secretary handle the matter as strictly an administrative affair, Haywood billed me an additional $1,400 for what he claimed were telephone calls and research into the matter. I had to converse with him by relaying messages through his secretary, and I quickly learned that we were paying $105 per hour for the services of his secretary. Haywood was billing for her work the full amount that he had quoted for his own work. I finally informed his secretary that Haywood had now been on the case for about seven months and had accomplished absolutely nothing; that I expected some type of response to my inquiry and some type of progress in the case. Haywood had not queried the District Attorney relative to the criminal aspect of the matter, nor had he ascertained the bank's position on failure to heed third party interest. He stated that his next step was to subpoena Mrs. Jaker to testify in the matter, and that he thought he could lead her into testimony which would involve the bank, or at the very least, establish her own liability in the matter. Haywood alibied that he had been terribly busy, but promised to have Mrs. Jaker come to his office to testify in about twenty days . Haywood claimed that he was ready to roll up his sleeves and start fighting; that he intended to start by issuing a subpoena to Mrs. Jaker. Immediately, I received another bill from Haywood. My wife and I anxiously awaited the results of Mrs. Jaker being subpoenaed. Two and one-half months passed and we still received no word from Haywood. I finally decided to call his secretary and inform her that if I didn't hear from Haywood, I was going to take him off the case. He still didn't return my call, so the next day I wrote him a letter notifying him that I was terminating his services. I told Haywood's secretary to please have the legal records in the case referred to my old attorney, David Hoffer. We had given up trying to sue the bank. David had no more answers than he had when we first employed him. He thought it would be better if we had Coward to obtain a final judgment, inasmuch as he was the attorney whose signature was on the preparatory documents. A few days later, Coward did the necessary work to obtain the judgment. Jaker did not contest the case. He had no intention of paying the judgment anyway. If the laws of the state and the United States permitted him to steal from his clients under the guise of his actions being civil not criminal, Jaker intended to take full advantage of such a bonanza. He was in fact "judgment proof". What few assets he had, he had already put into his wife's name, and thanks to Haywood Coward, Jaker had now learned that there was little likelihood of his being criminally prosecuted.

Paying off the bank loan. In mid-October, we paid the bank the $71,000 that Jaker had borrowed against our property. We also wrote checks paying approximately $6,000 to suppliers and subcontractors that Jaker had not paid. Jaker delivered lien waivers on the rest of the debts that he owed against the house. Succeeding events proved that the people who signed these waivers were cheated by Jaker when payment time came. With the house and lot returned to us, we knew that we had to proceed at full steam to get the house completed. The elements had been destroying the building and causing much work to have to be redone.

Protecting the property during the halt in construction. It was now Mid-October. We now had the property back in our names. The eighty day period during which construction had been halted had been one of the worst periods in our lives. Becky and I had worked constantly to prevent the elements from destroying the house. We had boarded the front and rear of the house to keep the rain from blowing in, and we eventually had a complete system of plastic basins to collect or divert the water coming through the roof. The 55 gallon drums that we had positioned throughout the house were filled to overflowing with each heavy rain, and the basement looked more and more like a deserted swimming pool that had never been drained. In the meantime, the partially completed house was being destroyed by thieves. My wife and I had worked in the August heat segregating and stacking thousands of board feet of lumber that Jaker's incompetent stewardship had permitted to be cut improperly and tossed into the front yard which had now grown into a huge junk pile of discarded Grade A lumber for which we had paid premium prices. Every morning when we arrived at the site, we would find that each stack of dimensioned lumber had diminished in size. The tire tracks of thieves who had stolen the lumber were invariably imprinted in the soft ground where they had loaded their trucks.

Apprehending a thief. One Sunday Morning, Becky and I drove to the site, and apprehended a man loading our 4' x 9' sheets of fir siding onto his truck. I confronted him and ordered him to unload the material. After identifying him by providing his automobile license number to the sheriff's department, we filed charges against him. The judge sentenced him to two years at hard labor which he suspended, based upon proper behavior and placed him on probation for five years, while fining him $250.00 plus court costs.

Staining the siding before putting it on the house. We were doing everything we could to lessen the expense, because we were now paying every debt on the house out of our personal bank account. My wife and I stained all the siding that was to be installed on the building, thereby avoiding the cost of staining after installation. We stained more than 100 sheets of 4' x 9' siding and stood it up to dry before restacking it. We also painted the inside of the building and we did all the caulking and as much of the carpentry as we felt that we were capable of doing.

Hiring a second builder. When my wife and I engaged our second builder, we wrote our own contract, carefully heeding all the things that we had learned from having been ill-advised by the lawyer who examined the first one. We engaged a young builder named Dale Cox who had built a house down the street. He agreed to provide his expertise and furnish the framing crew for a fee of $10,000. Under his contract, we paid all vendors and all subcontractors. The builder provided us with competitive bids which he received for all work on the house. He agreed to complete all the work within four months. Our biggest problem from that point forward would be with keeping track of the actions of subcontractors.

Dealing with subcontractors. When we were forced to start dealing with subcontractors, we encountered a variety of schemes commonly employed by them to increase their income at homebuilder's expense.

Many of them tried to take the shortest route to finishing the job. Many submitted inflated bills. Nearly all had mastered the art of "add-ons", a procedure whereby they claimed that they had done some type of additional work, and for which they added ridiculous amounts in the "total due" column. Some of these "add-ons" were in the range of fifty to three hundred dollars. There were others who exaggerated the quantity of material provided or the amount of labor performed. A major fiasco occurred when a roofing subcontractor and crew were left at the house on a Friday afternoon after my wife and I had gone home. The following morning when we returned to the site, we discovered that thieves had stolen the fifteen panel front door off its hinges and had taken twelve other windows and doors which had been inside the house awaiting installation, plus twenty sheets of siding valued at $25 per sheet. Altogether, the theft totaled nearly $3,000. The county detective agreed that the evidence was overwhelming; that the roofer was the guilty culprit, but he stated that we could not convict him unless we were able to locate the material which he had stolen. This theft brought the total dollar value of construction supplies stolen to $6,000.

Three successive sets of plumbers were necessary. Unskilled subordinates of the first plumbing subcontractor made a general mess of the entire operation. We paid the boss $1000 to give up the job and provide a lien waiver. The second plumbing subcontractor entrusted his work to a younger brother who was so unfamiliar with the trade that he would ask me what he was supposed to do next. We discharged this plumbing subcontractor also, but not before he had obtained approximately $1,800 in advance draw. The plumbing had already exceeded the fair value of what a subcontractor should charge, yet we had nothing to show for the funds that we had expended. When the plumbing was eventually finished, it had cost my wife and me approximately $2,000 more than it would have if a reliable subcontractor had been selected at the inception.

Two heating subcontractors were necessary. We had to discharge the first heating subcontractor because he absented himself continuously from the job site, even when his employees were making repeated errors. He failed to attend the job or to supervise the employees.

Dealing with inflated bills from subcontractors. When the tile installer arrived, I said to my wife, "I know what to expect, so I am going to prepare in advance. I am going to measure the area to be tiled right down to the square inch". My concern proved to be realistic, because the bill showed that he had laid 300 square feet of tile at $6.50 per square foot. I said, "Here are the precise measurements. You laid 246.5439 square feet.

That is a difference of $344.50." He grinned and said, "You'll have to admit I was pretty close". The insulation subcontractor inflated his bill approximately $180, only to correct it as soon as I showed him that his calculations were in error. The wallboard subcontractor turned out to be a piker. He inflated his bill only $90 on a $5,000 contract. He corrected his bill when I pointed out that he had charged for installing several sheets of wallboard that were still stacked in the basement.

Attempts to appropriate the builder's property. We were now hearing nothing from Beryl Jaker. We drove by his house occasionally and noticed that he had a new automobile and new truck parked outside the house. We ascertained that his new vehicles were registered in the name of his wife; however, he owned an old dilapidated car and truck, worth not more than $500. We reasoned that $500 was worth the effort, if we could appropriate the vehicles. We went to the sheriff's office and made a $50 deposit to defray the cost of a deputy visiting Jaker's home. When the deputy's report arrived, it was almost unbelievable. The deputy reported that the two vehicles were at Jaker's house, but they didn't have the license tags on them under which they were registered, and that he had been unable to trace the license tags; that consequently, he could not appropriate the vehicles. I asked, "Isn't it a legal offense to switch tags without notifying the law?" He replied, "Yes, but we don't do anything about cases like that. We went there only to appropriate the property, and since we couldn't positively identify the property, we left it alone."

Attempt to appropriate the property at the builder's new site. We decided to visit Jaker's lot in Bailey subdivision. We had learned that he had advertised the house at a price of $315,000. We made a special trip to the office of the Registrar of Deeds to research the files. We learned that the deed to the house and the mortgage loan was to Jaker and his wife and a Mohammed Komeni and wife. The first mortgage was in the amount of $183,000. We then reviewed the records in the Office of the Clerk of Court and found that Beryl Jaker and his wife filled two complete pages of the record. There were numerous liens and judgments against them ranging from liens and suits by lumber yards to various types of subcontractors who had performed work on his buildings and had been bilked. We learned also that an area Savings and Loan had a second mortgage in the amount of $22,500. A building supply company had a third mortgage in the amount of $36,000. Altogether, Jaker had already taken more than $240,000 out of the building. Conjecturing that Jaker's partner might be willing to negotiate some type of settlement, we went to the sheriff's office and exercised our right to make Jaker's interest the subject of a sheriff's sale. We had to deposit $500 to defray the cost of advertising and other miscellaneous expenses. We went to the court house on the day of the sale to see if anybody would be willing to bid on Jaker's house. There were too many liens against the property for us to bid on it and nobody else was gullible enough to bid on it. The house was deteriorating from exposure to the weather. My wife and I could not understand where Jaker was getting money to pay interest to the holder of the first mortgage. He was meeting his monthly interest payments which were in the vicinity of $1,500 per month which he paid on the $183,000 first mortgage for a period of two years before he defaulted. We learned of the foreclosure from a local realtor. Jaker was continuing to live in a new home that he had built with other people's money, and he was continuing to drive a new car and a new truck. However, we learned by the grapevine that the building supply company was foreclosing on Jaker's residence that he had occupied for two years without a certificate of occupancy. It was now obvious that Jaker had no intention of paying us what he owed.

Performing numerous construction tasks ourselves. On March 12th, we moved into our new home. Thanks to the determination of my wife and me, the house had taken on an entirely different look from the sad days when it appeared that it was going to be destroyed by the elements as a result of contractor negligence and misdeeds. The house now had distinct curb appeal and was just as beautiful as we had envisioned it before our heart wrenching experiences. For the first time, we were beginning to feel good about our efforts. It was almost as if the first rays of sunshine had broken through the clouds after a violent storm. There were a number of tasks remaining, but the outside of the house was beautiful. (See Figures 5 and 6). The eighteen hundred square foot basement had hardly been touched. We decided that we would finish it after we had lived in the building for awhile. We moved into the house exactly fourteen months after signing the flawed contract with Jaker.

Continuing our own construction after occupancy. We were glad to be rid of the construction workers, so we could assure a degree of care going into the work, which we planned to do ourselves. For the next several months, we busied ourselves applying a second coating of paint to the inside of the house, finishing the basement, and completing a large number of other tasks that we now felt certain that we were capable of performing. All of it together was work that would have cost us another $25,000 if we had hired subcontractors to do it, and it would not have been done as well as we did it. The house to this point had cost us $245,000, not counting the $31,000 that Jaker had stolen. We had to do all the finishing work that we could, for we didn't have the financial resources to hire it done. By the time we reached this stage, we were ready to attempt any task,

Figure 5: A view of the front of the house on our first day of occupancy

Figure 6: A view of the rear of the house on our first day of occupancy

regardless of its apparent degree of difficulty. We were convinced that we could do anything that the subcontractors could do, and in most case we could do it better and more economically. In reviewing the quantity and types of work that we did, it seems almost impossible that we did so many and varied tasks. We did all the cleaning of the house preparatory to painting, and we painted the interior and stained the exterior of the building. After the lawn's initial rough grading, we raked and seeded the one acre of lawn. My wife did all the landscaping. I built a brick sidewalk and small deck to complement the main deck of the house. We installed the kitchen chair railing and crown molding in the three rooms that had not been finished. We installed our own intercom system, and were joyfully surprised to find that it worked like a charm on the first try. Our son installed the garage door opener and the burglar alarm system. We removed a number of undesirable trees and replaced them with more exotic ones. My wife papered the kitchen, dining room, and five bath rooms, and the work was done without a flaw, far better than most of the wallpaper work that we have seen in the more expensive model homes. We became adept at a number of skills that we had never dreamed of developing. I became an accomplished carpenter and drywall man. Becky learned to stain and finish woodwork. I quickly mastered the art of doing our own electrical work. Had we not been thrust into an arena where we were almost compelled to do our own work, I am certain that we would never have attempted it. However, when we were compelled to do it, we did it, and we were proud of the results.

When we took final account of our nightmare, we figured that our persistence and determination had lessened the damages that had faced us at the beginning of the nightmare. In the early part of our effort, we had faced a potential loss of $73,500, a figure which included the $30,000 that we had advanced Jaker plus the $43,500 which we paid for the lot. By not throwing in the towel and by routinely assuming tasks which we had never dreamed that we could do, we managed to save our dream home, losing about $34,000 because we had engaged a general contractor who was devoid of any type of ethical considerations. Becky and I believe that if we had hired subcontractors to do all the work that we did we would have had to pay out about $35,000. In a sense, Jaker stole $34,000 from us, but we saved $35,000 by doing much of the work ourselves, so the building cost us just about what it would have cost if we had hired an ethical contractor and had done none of the work ourselves. By viewing our loss in this manner, we figure that the two year nightmare had cost us two years of worry and hard labor. On a further positive note, we attained an immense amount of knowledge and developed numerous skills that we would never have acquired otherwise.

THIRTEEN: PREPARING THE CONTRACT

By now you have become familiar with the several options that you have in contracting for the building of your home; i.e., doing your own subcontracting, contracting with a builder with the stipulation that you be permitted to perform selected building tasks yourself, or contracting with a builder to do one hundred percent of the work. It is assumed that by now, you have decided not to buy a ready built house; otherwise you would not have read this far. If you have elected to do your own subcontracting, the approach to the contracting effort will be different from the mode employed in contracting with a professional builder. Therefore this chapter discusses the procedures to employ in each case. The part which addresses subcontracting is limited to guidelines for the contracting that you will do with the subcontractors, inasmuch as the general details of subcontracting are discussed in Chapter 6. First, we shall consider procedures employed in acquiring subcontractors for the many tasks associated with the house.

Contracting with subcontractors. In Chapter 3, we learned that the technical name for subcontractors is mechanics; that they are legally empowered to place what is known as a mechanics lien against your house if they are not paid for their work. The term subcontractors came into use because they work for a general contractor, and in construction jargon it was necessary to distinguish between the general contractor and those who contracted with him for the performance of singular tasks. Strange as it may seem, there is seldom a formal document prepared when a subcontractor is engaged to work on a house. In ninety percent of cases, the agreement between a subcontractor and his employer is strictly verbal. During the building of my own home, there was not a single instance wherein the subcontractor asked for a written agreement. Admittedly, such oral agreements contain an element of danger. The courts have traditionally ruled that oral agreements resulting in exchanges of money, material, or services are binding contracts. The big difficulty stems from the necessity to prove what the agreement constituted, should such proof become necessary. However, the subcontractor task is generally easy to define. He is required to perform his tasks in accordance with the specifications. The plumber is required to plumb the house per specifications. The electrician is required to wire the house per specifications.

In like manner, other subcontractors are required to accomplish their tasks per specifications. If their work is not approved by the Building Inspector, they haven't complied with the code, and they are required to do whatever has to be done to obtain Building Inspector approval that the work is in conformity with the code. This does not necessarily mean that the work conforms with specifications. If you have doubts as to whether specifications have been met, you may want to consider obtaining outside evaluation assistance as outlined in Chapter 6. In most situations however, the only real source of potential disagreement between a builder and subcontractor is the proposed and accepted fee. If both parties have a dispute of the precise amount of money that was quoted by the subcontractor and accepted by the builder, there may be cause for seeking legal remedy.

To avoid such a situation, it is suggested that you have each subcontractor sign a statement whereby he agrees to perform his particular subcontracting task in accordance with the specifications for _____ dollars. (The blank to be filled in by the exact fee to which both parties agreed). If you require the subcontractor to perform work in addition to that spelled out in the specifications, you should take the necessary time to define clearly in writing the exact additional task work that you are requesting and the amount that you have agreed to pay for the additional work. You should then obtain his signature on the written document.

During the construction of my own house I had assumed that the rather large fee that my second builder had agreed to pay the trimmers included the trimming of the garage and construction of a stairway from the first floor to the basement. The trimmers argued that those items had not been considered when they placed their bid for the job. I had no choice but to pay them extra money for trimming the garage and constructing the stairwell.

You have already been alerted to the fact that many subcontractors are adept in the art of "add-ons", whereby they claim to have encountered something that they did not expect, or they have done something which the specifications did not require. In most localities, subcontractors are expected to call the Office of the Building Inspector when they have finished their work, and arrange for an inspection. The inspection fee is normally paid by the subcontractor.

Two admonitions. There are two more admonitions for dealing with subcontractors. The first is for you to recognize that many of them will request payment for their work or for a sizeable draw against their fee long before they finish the job. Be extremely careful in complying with these requests. If you pay too large a percentage of the fee at an early date, you may discover later that the subcontractor has moved to another job where the fee is more lucrative, and intends to finish your work when it is convenient for him to do so, thereby placing undue delay upon other tasks which cannot begin until his work is completed. The other admonition is to avoid to the extent practical the storing of expensive items of material at the job site at earlier dates than they will be required for installation in the building. When too many people know that there are merchandiseable items of material unguarded at a job site during the hours of darkness, the temptation and opportunity for theft are too great for some to resist. If you subcontract your own home, it is recommended that to the extent possible you be present at the job site each day as subcontractors leave the site. The important thing to consider is that you can save a relatively large amount of money by subcontracting your own home. To succeed in the endeavor, however, you must be able to spend at least thirty minutes per day at the job site to trouble shoot and to coordinate the work of the various subcontractors. You will also have to deal with the lending institution and the Building Inspector, and you must have insurance to protect you from any potential liabilities that may arise.

Do your homework before you even think about a contract. Preparing a contract for the building of your home is more difficult than any of the other options that you have considered. In the first place, you probably will be presented with a proposed contract that your chosen general contractor has prepared after querying you on the plans and specifications that you have provided him. Don't sign such a contract without first doing your own home work. When the time comes, you will want a competent attorney to review any contract that has been engendered by you, by the general contractor, or jointly between the two of you.

First, prepare a list of what you want in the contract. There are so many potential pitfalls in contracting for the building of a house that you should not accept a contract prepared by your attorney unless you have carefully prepared a list of all the things that you want in the contract, have provided the list to your attorney, and are satisfied that he has incorporated all your requirements into the contract. When most attorneys prepare contracts on their own with only cursory queries of their client, they refer to sample copies of contracts in their book of standard legal forms, insert a few of the suggestions that you may have made, and prepare the contract for yours and the general contractor's signature. Most contracts of this nature leave major issues unclearly defined and may or may not serve your best interest if the contract or any part of it should ever become the subject of a court case.

Insist on specificity. A contract should be an agreement specifying what, when, and where all aspects of the contract are to be carried out. The best thing to do is to define what is required so meticulously and in such great detail that there can be no possibility of the contract being misunderstood. It is not enough to say, "time is of the essence and, barring inclement weather, the contract will be completed four months from the date of signing this contract," or "the work will be of quality workmanship." That kind of wording is subjective, unmeasureable, and unquantifiable. What you want is a contract stating not only what is to be done, but when it is to be done, and that it is to be done in accordance with the specifications, including those for materials and grades of wallboard, roofing, and wiring, along with types and quality of plumbing with brand names. You should ask for a guarantee of all materials that are not under manufacturer warranty. The contract should contain a statement of who pays the fees of the county or municipality or utility companies and it should provide for a payment schedule, whereby you have the right of inspection before you pay. It should provide the precise date that construction is to begin and the date that it is to be completed with certificate of occupancy issued. You will identify the total cost of the contract effort, the hourly labor rate and any other debits which you will accrue. You will want the contract to establish specific milestones for completion of each phase of the project, with a monetary penalty involved if the milestone is achieved more than a week late. For instance, the contract should specify what date the footings are expected to be installed; the date the foundation is to be completed; the date that framing is to be completed; the completion dates of the plumbing system and the heating duct work, the roofing installation, the boxing, the setting of doors and windows, septic system, fireplaces, the insulation and dry wall, appliances installation, the carpentry and cabinetry, final leveling and grading, and of course, the date for final closing. The dates that you set for each milestone may be negotiated between you and the builder, but the results of your negotiation should be incorporated into the contract by your attorney.

Outline the construction tasks that you will perform. If you anticipate doing some of the work yourself, be certain that the tasks that you are to do are outlined in the finished contract. Check to make certain that the general contractor is not receiving pay for the tasks that you arc accomplishing. This may be the case if he has made his bid on the basis of the total square footage of the completed building.

Hook-up charges and money to be held in escrow. The contract should state what hookup charges are required for utility connections, and the amount of money that you are to withhold, pending your determination that the work has been done satisfactorily. If the builder argues this point, state that a specific sum is to be placed in escrow and released after all requirements have been met.

Make certain that the builder is bonded and that he has adequate workmen compensation. See that the contract specifies a requirement for proof that the builder has an adequate bond; that he has adequate workmen compensation and public utility. As an additional precaution, you should ask the contractor to show proof at periodic intervals that all bills attributable to the project have been paid, so they will not be charged to you at a later date.

Protect yourself against belated claims from subcontractors. You may want to consider the alternative of paying subcontractors yourself, so you can be absolutely certain that there will be no surprise liens placed against your house.

Obtaining Bids. After you have narrowed the potential contractor list to about four people as prescribed in Chapter 2, you should send a complete set of plans and specifications to each bidder. It is important that each bidder receive exactly the same package; otherwise you will have no really effective way of discriminating among the four bids. You should allow a minimum of ten days for them to come back with their estimates. You may want to allow them a full three weeks unless time is a vital consideration to you. When the general contractor provides you with a bid, he will identify all the tasks to be addressed along with his estimate of the cost of each task. He will provide you a separate list depicting his interpretation of the materials identified in the specifications, or his recommendation for materials necessary for changes or additional work upon which you and he may have agreed. This is an important list that must not be regarded lightly. Many general contractors use a standardized form for this purpose. The form is entitled "DESCRIPTION OF MATERIALS" and has been adopted by the Veterans Administration, the U.S.D.A. Farmers Home Administration, and the U.S. Department of Housing and Urban Development. These governmental activities have assigned the document form numbers as follows: HUD92005 (6-79); VA Form 26-1852;

and Form FmHA 424-2. All three forms are the same. They are used by the agencies for a variety of purposes such as HUD Application for Mortgage Insurance, VA Request for Determination of Reasonable Value, or FmHA Property Information and appraisal Report. Many general contractors use the form not only because it provides a ready format for identifying the description of materials going into a building, but because it also provides a ready made source of data in those cases where the data may be required by one of the listed government agencies. The form will be valuable to you because it provides you a detailed break-out of the many tasks necessary to complete a house and gives you a chance to solicit any outside assistance that you may need before discussing each item with your general contractor. The form may be obtained from the local U.S.D.A. FHA office, or from the nearest HUD or VA office. It provides an excellent step by step chronology in the nature of blanks that are to be completed jointly by you and the builder.

After you and the general contractor have tentatively agreed upon the DESCRIPTION OF MATERIALS, he may provide you a summation of the projected costs on a list similar to the following: (Costs quoted are examples only)

Note that the description includes an estimated fee of 15% for the builder's services. This is a conservative estimate. Chances are it will be somewhere in the range of 15% to 30% of the total costs of the finished building. After receiving this list, you should check as many of his estimates as time permits, so you can determine for yourself whether the estimates are realistic. Keep in mind that his total building fee is based upon the estimates of the costs of the various labor tasks and material that he has provided you. If his estimates appear to be realistic when you check them, use the same list to ask him to deduct the value of the work that you will perform.

You have already been warned to be especially wary of a contractor whose bid is considerably lower than the others. There are a number of factors which influence the amount of money that each contractor sets as his fee for doing the job. If he already has more work than he can perform expeditiously, he may make a high bid because he realizes that he would have to hire more general contractors and increase his own work load. He may make a low bid because he is about to run out of work and he needs an immediate job to keep his subcontractors employed, so he won't lose them. It may be a slow time of the year when construction has almost come to a halt. Keep in mind that a contractor who works on the job will generally make a lower bid than the office confined contractor who handles the work as an administrative matter.

Whatever you do, don't let the general contractor persuade you to award him a cost-plus-fixed-fee contract. If you remember, in Chapter 3 we

PROJECT_____   BUILDER_____
ARCHITECT_____   DATE_____ PHONE_____

BUILDING PERMIT AND INSURANCE	$1500.00
GRADING AND CLEARING LOT	1000.00
FINISH GRADING AND CLEARING DEBRIS	2000.00
LANDSCAPING	3000.00
WALKS AND DRIVEWAYS	2500.00
FOOTINGS	2375.00
BRICKWORK	13000.00
BRICK CLEANING	325,00
WATERPROOFING	450.00
BUILDING MATERIALS	53000.00
PEST CONTROL TREATMENT	100.00
FINISH CEMENT	1750.00
INSULATION LABOR	750.00
CARPENTER LABOR	18000.00
PAINTING	4200.00
WIRING	3500.00
PLUMBING	4000.00
HEATING AND AIR CONDITIONING	6000.00
SHINGLE LABOR NO.Squares__ x_$____per sq.___ =	7900.00
HANGING AND FILLING SHEETROCK	3522.00
GUTTERS	778.00
FLOOR COVERING	3952.00
HOOD, VENT HOOD, and VENT DRYER	150.00
LAYING FLOORS	1200.00
WATER AND SEWER	4500.00
APPLIANCES	2000.00
WALL PAPER	1000.00
MIRRORS AND BATH ACCESSORIES	500.00
SHOWER DOORS AND TUB ENCLOSURES	500.00
LIGHT FIXTURES	1500.00
WEATHER STRIPPING	100.00
TELEPHONES	100.00
OVERHEAD AND PROFIT 15%	22450.00
TOTAL	**$165,602.00**

Table 13-1 *Typical Contractor Estimate of cost of each task.*

labeled it as an instrument with the potential for financial destruction of the person contracting for the home to be built. You cannot depend upon the law to protect you from a dishonest contractor. The contract protects you only to the extent of the net worth that the contractor may have if things should go awry. If he has no net worth, you really have no protection. Keep in mind also that if the contractor represents a corporate enterprise and something should go awry, you would have to sue the corporation. You could not sue the contractor. If the corporation has no assets, your suit would be in vain.In Chapters 2 and 3, we discussed the characteristics that you should expect of a builder. In addition, you should check his credit rating and whether he has ever declared bankruptcy. It is important to know whether he has had liens or encumbrances filed against former houses that he has built. You can ascertain this by reviewing the records at the Office of the Clerk of Court. You may also check with the State Board of General Contractors to determine if he has a valid builder's license.

Include in your contract a procedure for any changes that you may decide to make during construction. Changes are where many building controversies arise. Many general contractors charge their clients thirty dollars for each change from the specification plus handsome amounts for labor and material necessary to effect each change. Your contract should specify when payments are to be made. Normally, a percentage of his pay is made at specific milestones as indicated above. It will be smart on your part to arrange to pay at each pay milestone that you designate, but pay a percentage of the total fee equaling only about one-half of the percentage of building completion at that specific milestone. In this manner, you can reserve a substantial amount of the contractor's fee until the job is 100% complete, thereby assuring the continued responsiveness of the contractor and giving you the necessary power to insist that all work be completed to your satisfaction.

The general contractor will probably ask you to deed your lot to him at the inception of the contract, if you own the lot on which he is to build. Don't comply with such a request. Remember that anything that you give the general contractor immediately becomes subrogated to the first mortgage that he obtains on the property, and if trouble should develop during the tenure of your contract, the bank may wind up with your money. If the general contractor argues that he is taking a risk by building a house on your lot, sell him the lot, but have your attorney arrange the sale in such a manner as to insure that the builder pays you before putting the lot up as collateral for a loan. The second request of the general contractor will probably be to ask you to make a down payment on the construction of the house. He may ask for as much as 10% to 20% of construction costs. You are advised to hold your ground and give him the minimum down payment possible. Inform him that you are willing to provide a minimum amount of earnest money pending his achievement of the first defined milestone, but you see no sense in your assuming the risk of something that is not under your control. If you agree to pay as much as $5,000, remember that it is at risk if for any reason, the bank should have to foreclose on the property. During the pre-contract negotiation, the question will probably arise as to your willingness to reimburse the general contractor for interest which he pays on the construction loan. This is not an unusual request, for the interest is either encompassed in the builder's overall fee or the buyer is expected to reimburse him for his interest expense. If you agree to reimburse him, make certain that he has obtained the loan at the lowest interest rate obtainable and be sure to place a cap on the amount of interest that you will pay, based upon the time quoted for completing the work. Obviously, you should not pay the builder's interest if the house goes several months beyond the projected completion date.

If you tie up any of your own money in the construction effort, as a down payment or otherwise, specify in the contract that the builder is to keep you continuously informed of amounts that he draws from the lending institution against the first mortgage. This action will keep you from being surprised if the builder withdraws funds in excess of his progress on the building and uses the funds elsewhere.

See that the contract states that the builder is not authorized to substitute one type of material for another, unless he first gains your approval and agrees to any monetary adjustments that may be in order as a result of the substitution. Have your attorney insert into the contract a statement that the costs of making corrections are not to be considered changes to the contract, and that such corrections are to be at builder's expense.

To make certain that every party to a contract knows and understands exactly what is expected, it is necessary to take the time to spell out all the details in writing. First you should write the specific work to be done in outline form, but make it detailed enough for your attorney and others to understand. The contract should contain cost figures for the job identifying when payments are to be made and in what amounts of money. For example, you might enter, "It is agreed that the builder shall receive a set fee of $30,000 to be paid as follows: 15% when the foundation is completed; 15% when the roofing sheathing is installed; 10% when wiring, heating, and plumbing have been roughed-in; 10% when sheet rock is hung; 20% when interior doors are hung and windows are trimmed; 10% when cabinets and tops are set; 10% after completion of landscaping and final seeding and raking of the lawn; 10% to be paid thirty days after all aspects of the contract have been completed to the client's satisfaction".

In addition to the above, your contract should state the exact date that the work is to be started and the date that it is to be completed. You will be smart to have your attorney insert a clause to the effect that for every day beyond the appointed completion date that the work has not been completed, the general contractor will forfeit one-half of one percent of his total fee for the contract.

You should include additional stipulations as to warranties and guarantees to which the general contractor agrees and any work for which the contractor is not going to be responsible, such as work that you may have agreed to do, or work for which you have agreed to be responsible. Don't fail to include the requirement for the contractor to clean-up the site after each work day, to include the removing of debris. Suggest to your attorney that he arrange to have your contractor post bond to protect you from any liens that might be filed by suppliers or any of the subcontractors. Also include in the contract the type, amount, and issuer of insurance coverage that the contractor carries. Include a three day cancellation clause in the contract. This will provide you three days to think about the contents before the contract becomes a final legal document.

If you have accepted the advice in Chapter 6 and have elected to hire a professional building inspector, a construction foreman, or a consultant to inspect the work as it progresses, this will be the time when you will realize the full benefit of whatever it has cost you. The inspector will have been checking the work on a weekly basis (more often when necessary). This will insure that both the contractor and subcontractors are putting out their best effort, because they will know that they aren't dealing with a novice in the construction industry. Even if they are so inclined, they will be reluctant to try to get by with shoddy work or poor quality materials. When the general contractor says that he has completed his work, you will want to call upon your inspector for a final evaluation and possible presentation of a "punch list" of things that the contractor will have to do before he gets paid for his work.

Last but not least is the requirement that the contract specify the conditions under which the contract may be terminated by either party.

If you heed the above guidelines, being careful to make a list of all the things that you want in the contract and take them to your attorney, you will wind up with a contract that will give you protection in the event the builder incurs problems that are beyond your control. Don't make the mistake that many people make when they naively permit their attorney to write a worthless contract. Outlined in the table below is a summary of the minimum elements of data that you should provide your attorney for incorporation into the contract for building your home.

Table 13-2 *Minimum Essential Details To Provide Your Attorney*

1. Detailed specifications prepared by the architect.
2. Details of any changes from plans or specifications.
3. Detailed description of materials for every task.
4. The need for a guarantee of all materials not under manufacturer warranty.
5. A statement of who pays the fees for county, city, or utility company services.
6. Your right to inspect and approve before you pay.
7. Date construction is to begin and date it is to be completed with certificate of occupancy issued.
8. Total costs of contract effort.
9. The hourly labor rate and any other debits that will accrue to you, the client.
10. Specific milestones for completion of each project phase, with a monetary penalty for each phase that is more than a week late.
11. Detailed description of any work for which you are going to assume responsibility, and the agreed monetary credit the contractor allows for your work, if any.
12. Identification of hook-up charges to be paid to utility companies.
13. Amount of money that you are to withhold pending determination that all work has been done satisfactorily.
14. Requirement for proof that the builder has an adequate bond and adequate workmen compensation.
15. Requirement for builder to show proof at periodic intervals that all bills against the building have been paid.
16. Name, address, and telephone number of the contractor.
17. State general contractor license number of the contractor.
18. Identification and copies of all warranties and guarantees.
19. A procedure for any changes that may be decided upon during construction.
20. A statement that correction of faulty work does not constitute change to the contract.
21. Statement of when payments are to be made; i.e., the percentage of pay that the contractor is due as he achieves each milestone.
22. The amount of down payment or earnest money that you are paying the general contractor.

23. The limiting date on any construction loan interest that you may have agreed to pay.
24. The maximum interest rate for which you are assuming responsibility.
25. A statement that the builder is not authorized to substitute one item of material for another, without your advance consent.
26. Identification of the type, amount, and issuer of insurance coverage carried by the contractor.
27. Insertion of a penalty clause for late completion of the work.
28. Insertion of a broom clean clause.
29. A three-day cancellation clause
30. Conditions under which either party may terminate the contract.

The next chapter addresses the pitfalls that may be encountered when one sets out to contract for the building of a home and identifies the ways that the pitfalls can be avoided.

FOURTEEN: AVOIDING THE PITFALLS

In the previous chapters we have identified several different options that you may pursue in acquiring the house of your dreams, and you have noted a number of different ways to save considerable sums of money without sacrificing quality. This book would not serve its purpose however, if it fails to acquaint you with the pitfalls to which you can be subjected if you fail to observe the fundamental principles contained in this book. You probably have recognized by now that the greatest potential pitfall of all is the possibility of selecting and entering into a contract with the wrong builder. Don't be frightened by the use of the term "pitfalls". In the final analysis, there are thousands of professional builders in the United States who are experts in their trade, who quote reasonable fees for their services, and who are conscientious about producing high quality work. Unfortunately, there are a number of builders who do not possess the characteristics that you want when you start to build your home. This book mentions many areas in which you can participate in the construction process, and/or in which you can assure yourself of a quality built home at reasonable price. None of the areas mentioned, however, is more important than the necessity to select a good general contractor. If you make that selection prudently and properly, you will eliminate nearly all the areas of concern that might otherwise be accompanied by stressful conditions and accompanying worry. Little more needs to be said about how to select a general contractor. That facet is adequately addressed in Chapter 2 and succeeding chapters.

In this chapter, we shall summarize some of the more important aspects that you have already read and, in addition, we shall point out helpful ways to avoid pitfalls that you could encounter if you haven't done your homework.

1. The poorly qualified builder. In the building profession as in other trades, there are a number of contract seekers who are either inept in their trade or intent upon taking advantage of their clients for purposes of self aggrandizement. In advising you of pitfalls that could be encountered, it is important for you to recognize that in your search, you probably will encounter at least one unprincipled builder. It will be helpful if you are armed with a few facts that will enable you to recognize him at the inception, so as to eliminate him from the selection process at the earliest possible time. He is not a lot different from other con artists, except for the fact that most con artists ply their trade by appealing to a sense of greed in their intended victim, often leading the intended victim to believe that he or she is going to get something for nothing. The greed tactic is certainly among the reservoir of tricks and schemes employed by the self seeking builder, but his primary mode of operating is to take advantage of the peculiarities of the legal system to cheat his client of large sums of money. Because his misdeeds are generally ruled as civil rather than criminal,an unethical builder is free to perpetrate his acts repeatedly within the same community. Since civil prosecution is the only thing about which he needs to worry, he protects himself by making certain that any property he owns is recorded in his wife's name.

A builder of this type generally is not native to the area in which he is operating. He has lived in his current habitat no more than three to five years; has been a party in several law suits in his former habitat; and at sometime during the past several years, he has declared bankruptcy. A check of the records in the office of the clerk of court will nearly always reveal that he has several judgments against him, along with numerous liens on property placed by vendors and subcontractors whom he has failed to pay. He may have been operating under several aliases, usually variations of his real name where first and middle names or initials have been reversed or modified with the last name remaining the same. The houses that he initially builds in a new area may or may not be built under contract. Frequently,he builds one or two modest houses as a means of acquainting himself with the various suppliers and subcontractors within an area, and of course, as a means of gaining the confidence of vendors and financial institutions with whom he hopes to deal in greater volume a little later.

Characteristically, he is glib of tongue. It is not uncommon for him to employ the smooth tongue technique to attract some individual more affluent than he as a partner in his business endeavor. The good name of the partner will be an important asset when time comes to borrow money from a lending institution. In much the same manner that a billiard hustler permits his opponent to win a few games, the builder permits his partner to share equally in early small endeavors, and continues this ploy until bigger money is at stake, at which time he is prepared to double cross his partner just as he would anyone else.

During the first two or three years that a fraudulent builder is in a new area, his every action is directed toward getting an unwary client's name on the dotted line of a contract . When a potential client presents himself, the builder employs his well developed spiel to convince the potential client that he builds better houses at far lower prices than any other building contractor in the area. He carefully explains that he is short of funds; that it will take some time for him to obtain the necessary construction loan; but, that he can begin construction immediately if the client will advance him fifteen percent of the contract value to defray expenses until the construction loan is processed. The builder counts heavily upon the fact that most of the attorneys in the area are unaware of his scheme and that, in most cases, they will advise the client that the contract is okay to sign. Almost invariably, his assumption proves to be correct. He does his best to con the unwary client into a COST-PLUS-FIXED FEE contract, for this is the type that gives him the greatest latitude for bilking the client; however, if he must settle for a fixed price contract, he will gladly do so. He knows at the time of signing exactly where he is to obtain the used or inferior material, and he has already planned a number of tricks of the trade to defraud the client. He makes a well organized appeal for an advance sum of money for any of a variety of reasons. A monetary advance is important to his scheme, because part of his plan is to build up total indebtedness against the house as rapidly as possible, so that no creditor can foreclose except the holder of the first mortgage.

If final settlement time arrives and he doesn't have the money to pay the liens, he simply leaves it to the client to decide whether to pay the liens or sacrifice whatever cash the client has put into the house. He doesn't care which alternative prevails. In the eyes of the law, he is simply an unfortunate general contractor who has tried his best and hasn't been able to make ends meet.

Once a self seeking builder obtains the contract and the advance of money, his fraudulent operation goes into full gear. He pays any of his old debts which may be causing him trouble. He upgrades his life style, and that of his family. He buys a new truck and new automobile. In the meantime, he is devoting no effort to the construction of the client's house. He figures that he has suckered the client into a contract and that the duped client doesn't know enough about construction to contest anything that he does. After four to six weeks the builder's construction loan has been approved with the client's lot now transferred to the builder, as security for the loan. The builder then begins to draw money against the construction loan, but he pays only about ten to fifteen percent of it to suppliers and vendors who have furnished supplies for,or worked on the client's house.

The builder has no intention of applying the money that he has drawn against the client's house to pay debts that have been accrued in constructing the client's house. The money is a far more valuable resource to the builder if he applies it on the start of another house which will generate another construction loan for him, and which will permit him to upgrade his life style even more. He knows that if he is careful not to record property in his own name, he can't be touched by the law. Knowing and depending upon these factors, the construction of the client's house proceeds at a snail's pace with vendor and subcontractor liens already beginning to accrue against the property for debts that the builder is evading in his determination to make maximum self use of the money that has become available to him as a result of having obtained the building contract.

In an effort to squeeze every dime out of his newly acquired opportunity, the crooked builder not only assures that construction proceeds at a snail's pace, he solicits vendors and subcontractors to over-bill for services rendered, in return for a partial rebate to them for their cooperation. If he is employed under a fixed price contract, he searches far and wide for used or inferior construction supplies to incorporate into the building, and he pays as few of the debts as possible, knowing that at the time of closing, the client must pay the bill, regardless of its size, even if there is twenty to thirty thousand dollars in liens against the building. He knows that if he doesn't have any property recorded in his name, he will emerge unscathed, and the client, who already has considerable money sunk into the building, will have to pay off the liens to avoid the possibility of losing his new home.

You should apply the above information when you start to consider potential candidates for the building contract. If you encounter one who portrays any of the characteristics identified above, eliminate him immediately from further consideration. There are too many good builders within the profession for you to take a chance on one who is either questionable of character or of talent, or of both. As a final admonition, don't make the mistake of "jumping into bed" with the lowest bidder, unless you use every guideline in this book to verify his credentials. Everybody looks for bargains. It is easy to understand why any of us would try to save several thousand dollars. Sometimes however, accepting the lowest bid does not save us money. More often than not it will cost money. If a contractor is worth his salt, he must make a profit. To do so, he has to pay his subcontractors and his suppliers. Then he has to pay for a certain amount of overhead. He can't possibly give you a quality product if he cuts his bid below a profit level. If you think that there is a possibility that a contractor is underbidding all others just to get the job, you should shun him like a plague. He intends to make a profit on your job, but it will be done in ways that could cost you enormous sums of money as well as any semblance of quality in your house.

2. Type of contract. There are two main types of building contracts, the Fixed price contract and the Cost-plus-fixed-fee contract. Under the fixed price contract, you, the client, agree to pay the general contractor a set price for a building that is to be constructed in accordance with the plans and specifications that you provide. Under a cost-plus-fixed-fee contract, you agree to assume responsibility for all costs associated with the construction, including but not limited to costs of material, labor, land, construction, loan interest, county and municipality fees, and all other costs, even those costs associated with correcting construction errors. In addition, you agree to pay the general contractor a set fee for his services. This fee may be expressed in terms of a specific number of dollars, or it may be expressed in terms of a percentage of the total costs of construction. The cost-plus-fixed-fee contract is a high risk contract for you, the client. Be extremely careful about selecting this type of contract. Make certain that you are aware of the risks before entering into it. These risks are identified in Chapter 3. More information can be obtained by consulting a competent attorney for consideration of the exact case. The fixed price contract also has certain inherent risks, as do all contracts, but the risks are not nearly as great with a fixed price contract as with a cost-plus fixed-fee contract.

3. Should you purchase a new home? Many people assume that contracting for the building of a home is too much of a hassle; hence, they spend their time and energy locating a new ready built home of their choice. Actually, contracting for the building of a home presents no more difficulty than purchasing one, provided you follow certain fundamental guidelines. The advantages you gain by having a home that coincides perfectly with everything that you ever wanted in a house are not the only advantages to be gained by going the contract route. First, there is the opportunity to save sizeable sums of money, and secondly, there is the comfort of knowing that the quality of the home will be substantially higher than the quality of a home that you buy on the market. In your contracted home, you need only follow a few practical and economic guidelines outlined in this book to assure quality. In a ready built home, there is no real way to assure quality. Too much of the construction is already covered with wall board and paint. Whether you buy a ready built home, manage your own subcontracting, or contract with a builder, the checks identified in Chapter 4 will prove valuable to you in evaluating a number of construction aspects to determine whether you are receiving a quality home.

4. Basic house construction terms. There are a number of basic house construction terms that will assist you in understanding how the construction process works. Like the skeleton of the human body, the real structural strength of any house rests in the foundation and the framing. The basic terms that will prove invaluable to you are defined in Chapter 5.

5. Subcontracting your own home. Managing your own subcontracting is not nearly as difficult as it may seem. Actually, it can be done by one who is a novice in the construction trade, provided he has the time to spend as much as thirty minutes per day at the construction site. If you need greater expertise than you think that you possess, you may hire all the assistance that you need for only a fraction of the amount that you would pay a general contractor. Tradesmen available for assisting you in this respect are builders whom you can hire as part time consultants. One also may hire inspector type foremen for as little as ten to twelve dollars per hour on a part time basis to assist in interpreting the plans, to help in acquiring subcontractors at reasonable rates, and to inspect the work of each subcontractor while construction is in progress. If you are willing to manage your own subcontracting, it is possible for you to save as much as $30,000 on a $200,000 home. Chapter 6 elaborates on the money saving opportunity.

6. Obtaining mortgage money. There are a variety of types of mortgages. The building of a home requires a construction loan mortgage and a permanent mortgage after the construction is completed. If you intend to manage your own subcontracting or for any reason, obtain the construction loan, you should attempt to obtain and process both loans simultaneously with the same lending institution. You do this to avoid double settlement costs. If you process the loans at separate times, either with the same or with different lending institutions, settlement could cost you as much as two or three thousand dollars each time. If both loans are processed at the inception, these settlement costs are paid only once.

There are three general types of loans, a Federal Housing Administration (FHA) loan, a Veterans Administration (VA) loan, and a Conventional loan. Any of the three may be acquired for a period of 15, 25, or 30 years depending upon the needs of the borrower. Other types of mortgages are the adjustable rate mortgages (ARMs), whereby the borrower agrees with the bank for the interest rate to be adjusted periodically to bring it in line with prevailing rates at the time. Another type loan is the graduated payment plan, whereby monthly payments are small for the first several years, but gradually escalating for a few years until it becomes fixed. The various type loans, along with some advantages and disadvantages of each are spelled out in Chapter 7.

7. Doing much of your own construction. You can save substantial amounts of money if you have the ability and the willingness to do much of your own work. Whether you manage your own subcontracting or engage a general contractor, there will be numerous areas where you or members of your family can complete tasks yourself. In some cases, it may be to your advantage to arrange to manage your own subcontracting for a few of the tasks associated with the project, while permitting the builder to complete the rest. It will be to your advantage to contract with a builder who is willing to grant you permission to do a number of tasks yourself. Not all builders will grant you the self help prerogative, but a lot of them will. The best source for contractors who are willing to grant you such permission is young contractors who have only recently acquired their builder's license and are anxious to make a successful start in the building profession. You are urged to check them out first for financial trustworthiness, but this can be done by following the guidelines in the early portions of this book. Chapter 9 provides a number of potential ways for you to do much of your own work.

8. The Building Inspector. The county or municipality Building Inspector performs a valuable service for one who manages the subcontracting of his home or engages a professional builder. Many builders depend entirely upon the Building Inspector for determination as to whether subcontractors have performed their work in conformity with the local code. In like manner he is a valuable resource for you if you haven't engaged professional help to inspect the work of the builder and his subcontractors. However, it is important for you to keep in mind the fact that the Building Inspector is required to inspect only for conformity with the code, he does not always inspect to assure conformity with specifications. If the specifications happen to exceed code requirements and you don't feel capable of performing the final inspection yourself, it would be wise for you to think about employing a paid inspector to make the determination of whether the specifications have been met.

9. Tips for saving money. Chapter 10 provides twenty-six tips for saving money. They are practical realistic things that almost anyone can do if he has the time to devote to the project. It is possible to reduce the cost of your home by as much as twenty-five percent by incorporating less expensive material in areas where it doesn't make a lot of difference, by doing a lot of the labor yourself, or engaging a general contractor only for selected basic framework and the completion of only one or two rooms, while reserving the rest of the effort for you to do yourself at a later date. You don't have to be licensed for work that you do in your own home; however, you do have to perform the work in such a manner that it meets the requirements of the local code. This is really not difficult to do because the Building Inspector will tell you what is wrong if an item of work fails inspection, and will tell you exactly what you have to do to correct any deficiency.

      10. Preparing the contract. There are approximately thirty important elements of information that should go into any contract for the building of a house. All these elements are listed in Chapter 11. It is up to you the client to identify the things that you want your attorney to insert into the contract. It is then the job of the attorney to put your information into proper legal format and to advise you if there is any additional item that should be addressed. Selecting a good contractor to build your house is the most important aspect of contracting for the building of a home, but wording the contract and making certain that it contains all the essential elements is the second most important step. You should use the information contained in Chapter 11 as a check list to see if you have provided your attorney with all the necessary information. Every contract must take cognizance of some variables or unpredictables that may arise during the time that the contract is in force. You can request that these elements be excluded from the contract, or you can have your attorney to write them in as allowances. Each allowance would specify an amount of money to be set aside for a particular job or material;that if that item costs more than anticipated the client is to pay the difference. If it costs less, the contractor is to credit you with the difference.

      11. Executing the contract. Once the contract has been signed, the builder starts executing the many varied tasks to be performed. Hopefully, you will have selected a good contractor and that as a result of your having done so, your participation in the effort can be reduced to the minimum time that is consistent with your desires. However, regardless of the proficiency demonstrated by the builder, you will still want to assure yourself that you are receiving the type of building that you envisioned when you first selected your plans. You may or may not have negotiated a contract which provided some flexibility for you to perform or supervise the performance of specific portions of the overall effort. Regardless of the degree of your own participation, it will be smart of you to engage a consultant or paid inspector to inspect any work that you do not understand or do not feel capable of adjudging for yourself.

      Consultants and paid inspectors hired on an hourly basis are relatively inexpensive when compared to their potential benefit to you in consideration of the total costs of the overall project.

THE END

INDEX

ABOUT THE AUTHOR

Gus Nance is a retired lieutenant colonel, US Army Corps of Engineers and also a retired federal civil servant from the US Army Materiel Command. Gus served for approximately twelve years in facilities engineering assignments where he was responsible not only for maintenance and repair of facilities, but also for inspection and approval of new construction within his territorial jurisdiction. Toward the end of World War II, he gained early construction experience when as Commander of an engineer construction unit, he was responsible for erecting new buildings which constituted the early framework of the University of the Philippines. Later, he supervised contract performance in the construction of the Norristown Armory, near Philadelphia, Pennsylvania. In the 1950s, he served as engineer operations officer at Camp Drum, New York where he supervised the design and construction of roads, bridges, and range facilities to accommodate a continuous summertime presence of a national guard and an army reserve division. In Greenland, in the early 1960s, he monitored thirty R & D projects for the Chief of Engineers, several of which related to development of shelters for operations in the Arctic. As a federal civil servant, Gus served in positions where he was responsible for the formulation and publication of new doctrine for army combat units. He spent approximately five years writing doctrine for the combat arms and approximately five years for the development of materiel. He authored numerous technical and field manuals and was also the chief of study teams performing in-depth studies to develop doctrine in step with the advances of technology. Among these studies were the US Army Prisoner of War Study and the Civil Military Operations Study. Gus also authored the 200 page technical publication, *The US Army Combat Developments Command Methodology Notebook for Action Officers.*

As a consultant, Gus has written several successful contract proposals for industry. His writing of *CONTRACTING TO BUILD YOUR HOME* came as a result of his personal experiences which reflected a vital need for analysis, documentation, and publication. After a builder defaulted on the contract for Gus' own 5000 square foot residence, he discharged the contractor and assumed personal responsibility for completion of the last seventy-five percent of the construction work, correcting faulty work of the errant builder in the process.